A communication system must have adequate electrical protection in order to meet the reliability standards for commercially acceptable service and to keep down maintenance expenses. This report characterizes, from the viewpoint of electrical protection and coordination, the conditions of the electrical environment to which communication facilities are exposed. It gives consideration to both present and anticipated future conditions, and covers such topics as the effects of lightning, interference from power networks, electric shock, earth potential gradients, corrosion, over-voltage in AC power utilization circuits, and electromagnetic pulses.

Characterization of the Electrical Environment is a current reference on the design factors required to ensure reliable performance of communication facilities under field operating conditions. It will be useful as a manual for practising engineers in telecommunications, and as a tutorial textbook in engineering schools in North America, Europe, and elsewhere.

It was originally issued as a Systems Engineering Report by Bell-Northern Research Ltd., Ottawa.

CHARACTERIZATION OF THE ELECTRICAL ENVIRONMENT

David W. Bodle
Axel J. Ghazi
Moinuddin Syed
Ralph L. Woodside

University of Toronto Press
Toronto and Buffalo

© University of Toronto Press 1976
Toronto and Buffalo
Printed in Canada

ISBN 0-8020-2194-8
LC 76-22886

CONTENTS

PREFACE

Communication facilities are normally exposed to a hostile electrical environment, especially in rural areas. It is therefore necessary to study the environmental hazards and to characterize exposure conditions in terms of system design criteria so that the type and degree of protection required by various facilities may be determined.

The purpose of this document is to quantize the present and future environmental exposure factors affecting communications systems. This will provide protection, system, design, and development engineers with the best current reference source covering the electrical protection design factors required to ensure reliable performance of communication facilities under field operating conditions.

The environmental factors that an engineer faces are strongly influenced by the atmospheric conditions, the soil, the topography, and the effect of other utilities in the particular location under study. These influences must be understood, evaluated, and properly accounted for in all electrical protection systems. Because of the variable nature of these influences, relevant protection design information is presented on a probability basis whenever possible. As well-defined areas of information are quite limited, the values offered as design guidelines in some cases have been derived from the best data available and incorporate judgement based on field experience. There are also a few subjects about which very little quantitative information is available. These voids are pointed out and clearly indicate fertile areas for future investigation.

Electrical protection has at times been regarded as little more than a 'necessary evil', probably because it does not produce revenue directly. However, there is no basis for such an attitude, since without adequate protection a communication system would not meet the reliability standards for commercially acceptable service and maintenance expenses would become extremely high.

This document was originally published in 1973 as a Systems Engineering Report (SER 156) by Bell-Northern Research.

Reference has been made to material obtained from the American Telephone & Telegraph Co., and Bell Canada, Ontario Hydro and Hydro Quebec, in several sections of this report. Some of this material is of a proprietary nature and may have restricted circulation.

ACKNOWLEDGEMENTS

Those responsible for preparing this report wish to express their appreciation for the contributions and assistance received from other groups.

This includes the representatives from Ontario Hydro and Hydro Quebec who have provided both information on their present operations and views on the future power network.

Assistance has been provided by Mr. R.A. Conley, Mr. V.B. Pike, and other members of Bell-Northern Research.

The information on the Bell Canada Network, its operation and other contributions were provided by Mr. B.C. Nowlan, Mr. P. Ferland, and Mr. J. Solinas.

1. INTRODUCTION

1.1 PURPOSE

The purpose of this document is to characterize the electrical environmental conditions to which communication facilities are exposed. Consideration is given both to present electrical environmental conditions and to anticipated future conditions. The environmental information which is presented is intended to facilitate the development of system protection requirements. These requirements will be determined after input is analysed from all of the other related factors as shown in Figure 1-1.

Environmental information presently available adequately defines many of the areas of principal concern; however, critical voids do exist, particularly in areas of a quantitative nature. These less defined areas are identified and examined in the document so that further investigation can be most profitably directed.

1.2 SCOPE

1.2.1 Protection versus Noise Domain

The document deals with those aspects of the electrical interference problem that are commonly recognized to be within the 'protection domain', namely, extraneous voltages and currents that constitute a shock hazard or may damage facilities. Disturbances that merely interfere with the quality of transmission are customarily referred to as 'noise' and will not be dealt with in this text.

In considering the respective domains of noise and protection, it is interesting to note that in the past all circuit components had substantial dielectric strengths and current carrying capacity that permitted a rather simple method of distinguishing between noise and electrical protection problems, i.e., disturbing voltages that did not constitute a shock hazard or would not damage facilities were considered to be in the noise domain. This was the period when 3-mil carbon protector blocks having a nominal spark-over voltage of 500 peak volts dictated equipment dielectric design levels. Design standards have changed due to miniaturization and the use of semiconductor circuitry, and equipment may now be subject to damage by extraneous potentials of only a few volts. Therefore, noise and electrical protection are now frequently concurrent problems.

A glossary of terms commonly employed in the electrical protection field is provided in subsection 1.3.

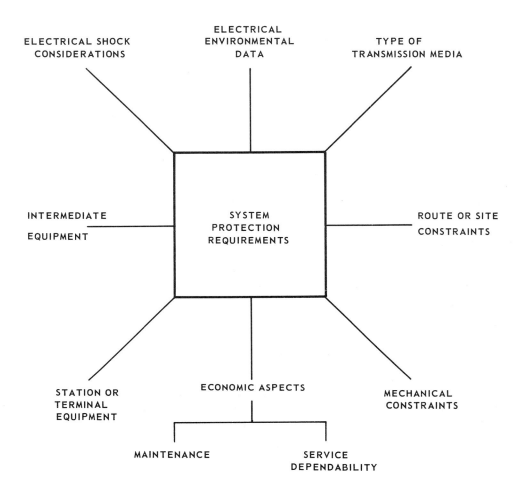

FIGURE 1-1

Inputs for Determining System Protection Requirements

1.2.2 Areas of Usefulness

Characterizations of extraneous voltages, currents, and wave-shapes, as well as other environmental information, have the following applications.

a) Development of system protection requirements.

b) Development of design test standards for system components.

c) Design of protection devices — development of test specifications.

d) Cable design — basis for determining dielectric strength, shielding, and corrosion requirements.

e) Equipment design — basis for determining protection measures and required dielectric strength of components.

f) Grounding requirements and protection of structures at equipment locations, ranging from manhole repeaters to customer station, PBX, central offices, and radio systems.

g) System application problems — provide direction in determining transmission media; also aid in route and site selection.

1.2.3 General Protection Considerations

This document does not present detailed information concerning protection devices and their application, since these aspects are covered in other reference material. However, the broad procedural aspects of protection engineering will be reviewed briefly to provide some initial perspective.

It is a generally accepted fact that 'tailoring' protection to meet the specialized requirements of many types of system components would lead to serious administrative difficulties. The established practice, therefore, is to standardize protection devices and measures on the basis of rather broad and general types of plant and equipment. An example of such standardization is provided in Table 1-1.

TABLE 1-1

Standardized Protection Devices

GENERIC CLASSIFICATION	BASIC PROTECTION	
	TYPES OF PROTECTION DEVICES	NOMINAL PROTECTION LEVELS PROVIDED
Customer Station Apparatus (Station sets, Key sets, PBX)	3-mil carbon protector blocks in fused or fuse-less mountings	Lightning Surges: 800 pk V 60 Hz Induction 425 V rms
Central Office and other equipment centers	Main frame protectors Heat coils	Lightning Surges: 800 pk V 60 Hz Induction 425 V rms 0.35 A (sneak current)
Cable plant such as at junctions with wire plant, and at paper-PIC cable junctions	6-mil carbon protector blocks	Lightning Surges: 1200 pk V 60 Hz Induction 700 V rms

The protection devices listed in Table 1-1 provide 'basic protection'. They place an initial ceiling on extraneous voltages and thus define the first level of exposure to which associated equipment may be subjected. The protection levels provided by discharge type devices (carbon blocks or gas tubes) are usually adequate for circuit wiring, electromechanical devices, capacitors, and other common circuit components. The effectiveness of protectors depends to a considerable extent upon the manner in which they are connected to the exposed circuit. The sparkover values given in the table are those that appear at the terminals of the protector mounting — long connection leads can seriously reduce this effectiveness.

'Supplementary protection' is additional protection required for semiconductor components and microcircuitry. This protection frequently takes the form of zener diodes to limit voltage and additional resistance in series with critical transistor junctions to limit current.

Much of the information presented in this document is available to anyone prepared to devote considerable time and effort to search the literature. Only those data that, in the opinion of the editors, are directly relevant to the development of system protection requirements have been abstracted. The more relevant published sources of information have been referenced to facilitate further study if required.

It is anticipated that the concise manner of presentation and the establishment of design test standards will ease the task of designers of telecommunication equipment.

1.3   GLOSSARY

Arrester
A protection device used on power lines to limit line-to-ground surge voltage while simultaneously preventing 'power follow' current (i.e., the grounding of normal power).

BIL
Basic Impulse Insulation Levels (BIL) is a reference impulse insulation strength expressed in terms of the crest value of withstand voltage of a standard impulse voltage wave.

Cable
Transmission path composed of a *core* of numerous paired or quadded wire conductors, coaxial tubes, or any combination thereof. Conductors are separated from each other by insulation and covered by a *sheath*. The sheath is generally available in a variety of combinations of layers of polyethylene, armor wire, or tape, and usually includes a solid, continuous metallic *shield*.

Cable, Aerial
Cable designed to be installed on aerial supporting structures such as poles, sides of buildings, and other structures.

Cable, Buried
Cable designed to be installed in the earth in direct contact with the soil.

Cable, Underground
Cable installed below the surface of the earth in utility tunnels, manholes, or inside conduit made of metal, concrete, tile, fiber, plastic, or any other composition material.

Core
    See Cable.

Corona
    An electrical discharge resulting from ionization produced
    by a strong electric field. Corona manifests its presence
    in the form of a glow that is visible in the dark, often
    accompanied by a hissing or buzzing sound.

Earth Currents
    Any alternating or direct currents, whether momentary, inter-
    mittent, or continuous, that flow in the earth. These
    currents may be caused by power systems, power faults to
    ground, lightning, solar phenomena, geophysical magnetic dis-
    turbances, or nuclear explosions (see Voltage Gradient and
    Earth Potential Gradient).

Earth Potential Gradient
    The change in voltage per unit distance in the earth due to
    earth resistivity; the voltage is caused by the man-made
    conditions or natural phenomena listed under Earth Currents
    (see also Voltage Gradient).

Electric Induction
    A process by which voltages are induced in a conductor, such
    as a telephone line, by the electric field of nearby parallel
    charged conductor(s), such as power transmission line.

Environment of Telephone Plant

    Metropolitan or Urban
        Heavily built-up areas where plant is chiefly underground
        or shielded from direct lightning strokes by tall
        structures. Induction in cable pairs is low because of
        the substantial shielding provided by bonding the shields
        of cables in multiple cable runs and by the presence of
        other buried metallic systems. Such areas are typically
        classified as areas of limited exposure.

    Suburban
        Built-up areas involving primarily residential structures and
        scattered tall structures where the telephone plant is most-
        ly aerial and buried and may be classified as exposed or
        minimally exposed by designated blocks or wider areas (see
        Cable, Aerial and Cable, Buried). Where there are three or
        more metallic-shielded cables in a common underground run,
        the exposure of the pairs will generally be moderate.

Rural

Sparsely settled areas in mostly open country where the telephone plant is both aerial and buried and is usually classified as exposed.

Exchange Cable

Any one of several types of high-capacitance ('high cap') cable plant used primarily in metropolitan or suburban areas (q.v.) for interoffice trunks and subscriber loops. High-capacitance cable is not normally used in trunk cable plant above end offices (class 5 offices) where greater distances are involved (see Trunk Cable).

Follow-Current

Most frequently used with regard to the current from power sources that flows through protection devices (lightning arresters and protectors) following the passage of the initial surge.

Ground (Effectively Grounded)

A conducting connection, whether intentional or accidental, by which an electric circuit or equipment is connected to the earth or to some conducting body of relatively large extent that serves in place of the earth. It is used for establishing and maintaining the potential of the earth (or of the conducting body) at approximately the same potential as conductors connected to it and for conducting ground current to and from earth (or the conducting body). A system or device is *effectively grounded* when it is permanently connected to earth (or the conducting body) through a ground connection of sufficiently low impedance and of sufficient ampere capacity to prevent the buildup of voltages that may result in undue hazard to persons or connected equipment.

Hardening

The modification of structural plant and equipment design to facilitate resistance to the blast and shock effects of an explosion — usually nuclear — and the shielding and protection of plant and equipment circuitry from the radiation effects and electromagnetic pulse (EMP) transient effects that accompany a nuclear explosion.

Impulse

A surge with unidirectional polarity, having a short time duration.

Induction

See Electric Induction and Magnetic Induction.

Inductive Coordination

The cooperative efforts of telephone and power company personnel in the engineering and application of protective measures and devices in arrangements of telephone and power networks to obtain optimum results.

Isokeraunic Map

A geographical map of a wide area, such as Canada or the
United States, showing continuous lines connecting points of
equal *thunderstorm-day* activity. From the data presented, a
relative comparison can be made of the thunderstorm activity
in one area with that of another area.

Longitudinal (Common Mode)

A mode of voltage, usually with respect to ground, and
resulting current (known as longitudinal current), which flows
in the same direction, along associated conductors (pair or
pairs) having a potential with respect to ground. This
current uses the earth or other grounded conductors as a
'sink' or return path as appropriate.

Loop

The wire pair that connects the subscriber to the end
(central) office. This path is called a loop because it is
electrically and physically described on the basis of the
round-trip distance from office to subscriber station and
return.

Magnetic Induction

A process by which currents are induced in a conductor, such
as a telephone line, by a magnetic field from nearby parallel current carrying conductor(s), such as a power transmission
line.

Metallic (Transverse or Differential)

A term used to describe voltage producing currents of equal
magnitude that propagate in opposite directions in the two
wires of a pair. Such currents are also referred to as metallic, transverse, or differential.

Ohm-Metre

A unit of measurement commonly used in engineering literature
to indicate the volume resistivity of a cubic metre of
material (usually soil). With regard to soil, it is the dc
resistance obtained with electrodes placed against opposite
faces of a cubic metre of soil.

Note: Geophysicists tend to express the resistivity of
soil in ohms-centimetre, which may be converted to
ohm-metre by dividing by 100.

Primary Exposure

The initial disturbing source, such as lightning stroke
current or a 60 Hz primary field.

Primary (Disturbing) Field

As used in low frequency induction problems, it is the voltage
induced into a disturbed circuit due to current in a disturbing

circuit in the absence of current in a shield circuit. It is synonymous with nonshielded induced voltage.

Protection

Employment of measures and/or devices to limit voltages and currents to values that will not constitute a shock hazard or damage the components of a communication system.

Protection, Basic

Fundamental protection measures (e.g., grounding and shielding). Also heavy duty devices such as carbon blocks or gas tube protectors applied directly to transmission media to provide initial voltage or current limitation.

Protection, Supplementary

Measures and/or devices that are used in conjunction with basic protection to further limit disturbing voltages. This protection frequently is in the form of a current limiting impedance and low voltage, often by use of moderate-duty semiconductor devices such as diodes or four layer devices.

Protector

A protection device used on communication systems to limit excessive voltages. Such a device may consist of closely spaced carbon electrodes discharging in air, or gaps in an envelope containing gas at a reduced pressure.

Random Separation

The practice of burying communication and power supply cables at the same depth with no specific separation between facilities. For telephone plant to be so constructed, certain conditions and requirements must be met in connection with supply voltage limitation and the use of bare grounded conductors in various forms.

Soil Resistivity

The measured dc resistance of a unit volume of soil (earth) expressed in proper units.

Stroke Factor

The number of lightning strokes to earth per square mile per thunderstorm day.

Terrain Factor

A quantity, used in estimating lightning stroke incidence to plant, to compensate for deviations from average (unity) exposure conditions due to variations in features of the terrain.

Thunderstorm Day

Any day during which thunder is heard at a specific observation point.

Note: The recognition of one thunder clap during the period is sufficient to classify it as a thunderstorm day. Consequently, such observations merely confirm the presence of lightning but provide no information on the number of strokes to earth.

Trunk (Toll) Cable

Any of several types of low-capacitance ('low-cap') cable plant used to provide trunk connections. Such cables are ordinarily of substantially longer length than interoffice trunks.

Virtual Zero (Impulse)

The virtual zero point of an impulse in a conductor is the intersection with the zero axis of a straight line drawn through points on the front of a *current* wave figure at the 10 percent and 90 percent crest values, or through points of a *voltage* wave at the 30 and 90 percent crest values.

Voltage Gradient

The change in voltage per unit distance in a substance (e.g., in a length of conductor or a volume of soil).

Wave (Electric Circuit)

The variation with time of current or voltage at any point in an electric circuit.

Wavefront (Impulse)

That part of a wave between the virtual zero point and the point at which the impulse reaches its crest value.

Wavetail (Impulse)

That part of a wave between the point of crest value and the end of the impulse.

Waveshape of an Impulse (Standard Designation)

The waveshape of an impulse (other than rectangular) of voltage or current is designated by a combination of numbers. The first, an index of the wavefront, is the virtual duration of the wavefront in microseconds. The second, an index of the wavetail, is the time in microseconds from virtual zero to the instant at which one-half of crest value is reached on the wavetail (e.g., 1.2 × 50 or 10 × 1000).

2. LIGHTNING EFFECTS ON COMMUNICATION SYSTEMS

2.1 FUNDAMENTAL ASPECTS

2.1.1 General

Lightning is a major environmental hazard to communication systems and will be covered in detail in this section. The material is presented from an engineering standpoint and has been selected on the basis of its usefulness to the protection engineer. Attention is directed exclusively to the effects of lightning on land-based facilities. Those interested in such aspects of the subject as cloud formation, separation of charge, and other atmospheric details should consult other texts.[1]

Lightning is an electrical discharge from cloud to cloud or between cloud and earth. Cloud-to-cloud discharges are by far the more numerous but are of minor concern to the engineer responsible for the protection of subscribers, plant personnel, and plant equipment. Electrostatic induction in communication plant produced by cloud-to-cloud discharges is of relatively low magnitude, and although such induction is a source of 'noise', it is rarely a hazard to personnel or insulation.

Lightning strokes to ground are the greatest single source of hazardous foreign potentials and currents in communication plant. Direct strokes to wire and cable may cause serious arcing at the point of contact and, since the plant becomes a part of the series path between cloud and earth for large-amplitude surge currents, voltage differences develop in the plant that are of sufficient magnitude to cause dielectric breakdown and produce hazardous shock conditions. Strokes to earth near communication plant may not involve the plant directly through a *conductive* discharge path, but fields associated with such strokes may be of sufficient intensity to produce *inductive* effects on personnel and plant.

Lightning occurs in practically all parts of the world, but the yearly incidence of thunderstorms and their relative intensities vary greatly with location. It must be assumed, therefore, that plant will be exposed to the damaging effects of lightning to varying degrees. In rural areas, the exposure will be greatest because little benefit can be expected from shielding provided by other grounded structures. Plant located in cities and other built-up areas benefits from the presence of high structures that intercept strokes, and from shielding provided by other grounded conducting media, such as public water systems.

The source mechanism of lightning strokes to objects on the earth is well enough understood for engineering purposes. Quantitative values of fundamental parameters are available in the reference literature. Much engineering information has been obtained from measurements on power transmission lines, and probability curves may be found in reference books and transaction papers concerning such aspects as current magnitudes, waveshapes, coulomb content, and multiple stroke components. Data of an approximate nature are also available with regard to the distances over which strokes to earth are likely to be diverted to aerial lines and structures. Primarily, the susceptibility of an object to direct lightning strokes is a function of its height, the polarity of the thundercloud, and the incidence of thunderstorm days.

Referenced data offer a way of roughly predicting the probability of strokes to communication plant, e.g., aerial lines, cables, and antenna support structures. Predictions of this nature are based on average conditions and so can be subject to considerable error when local conditions vary from the mean. Stroke incidence and intensity may vary widely from year to year as illustrated in the Empire State Studies[8]: although the average yearly stroke incidence was 23, yearly variations from 2 to 47 were observed over the period of the study. It must be recognized that probability figures are only valid when related to a significant period of time, and the above example should suggest the inadvisability of placing great dependence on short time observations.

An effort has been made to define environmental conditions on a probability basis, since this provides the most effective means of establishing system protection requirements that optimize cost-benefit relationships.

2.1.2 Types and Incidence of Thunderstorms

Thunderstorms are of two general types:

1) convection storms, which are local in extent and of relatively short duration, and

2) frontal storms, which extend over greater areas and may continue for long periods of time.

Convection-type thunderstorms account for the majority of annual thunderstorm days in the United States and Canada, but studies and experience have shown that a frontal-type storm will cause appreciably more damage than a convection-type storm. Convection-type storms are caused by local heating of the air near the earth; consequently these storms predominate during the summer months and in the warmer climates. They are nonregenerative in nature because rain accompanying such disturbances soon cools the earth and dissipates their source of energy. Limited observations indicate that the magnitude and especially the incidence of lightning strokes to ground in such storms are materially less than in frontal-type storms.

Frontal-type thunderstorms result from the meeting of a warm, moist front with a cold front, and at times may extend for several hundred miles, exposing large areas to severe lightning discharges for several hours. These storms are regenerative in nature, because the air masses may continue to move in and maintain the turbulence. Extensive data on the annual incidence of thunderstorm days have been compiled for most areas of the world. These data have been plotted in the form of *isokeraunic* maps (Figures 2-1, 2-2, 2-3), which are used extensively for estimating plant exposure to lightning.

2.1.3 Stroke Incidence

A *thunderstorm day* by definition is any day during which thunder is heard at a specific observation point. These observations merely confirm the presence of lightning and do not provide the information on the number of strokes to earth that is ultimately required for estimating probable plant exposure. However, through field investigations[2] conducted in North America, statistical data suitable for engineering purposes have been secured on the number of strokes to ground per square mile per thunderstorm day. These values are referred to as *stroke factors*, and are as follows:

a) In areas experiencing a relatively large number of frontal-type storms, the stroke factor is 0.37.

b) In areas where most storms are of the convection type, the stroke factor is 0.28.

2.1.4 Thunderstorm Electrification

The type of cloud formation normally associated with lightning generation in either convection or frontal type storms is the *cumulonimbus* cloud, more commonly referred to as the 'thunderhead'. Strong winds, rain, hail, and lightning result from such clouds, which, when built up fully and viewed from a considerable distance, have a characteristic anvil shape with a broad top, a sharply defined outline, a brilliant white appearance toward the top, and an ocasional fleecy cap of ice crystal. Thunderstorm research has led to several theories of electrification that are deficient, or at most marginal, in accounting for the quantity of charge production required to produce lightning. In the last decade, however, considerable progress has been made toward understanding this electrification process and developing a model for thunderstorm electrification[3].

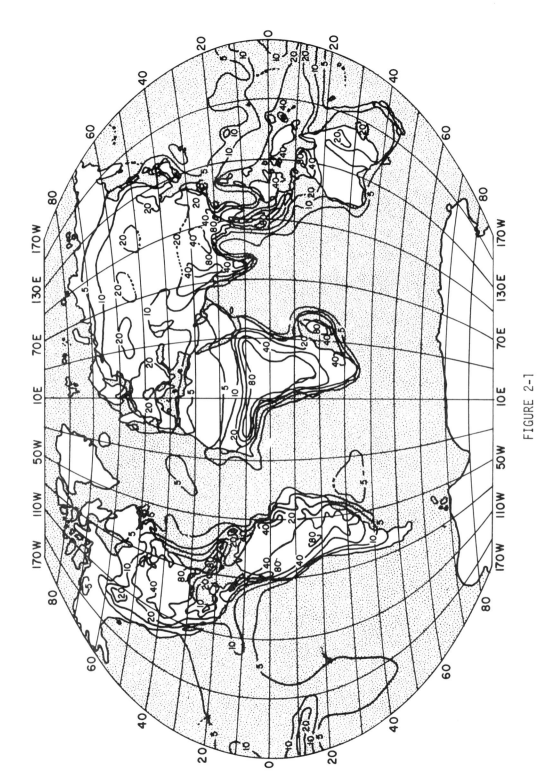

FIGURE 2-1

World Isokeraunic Map

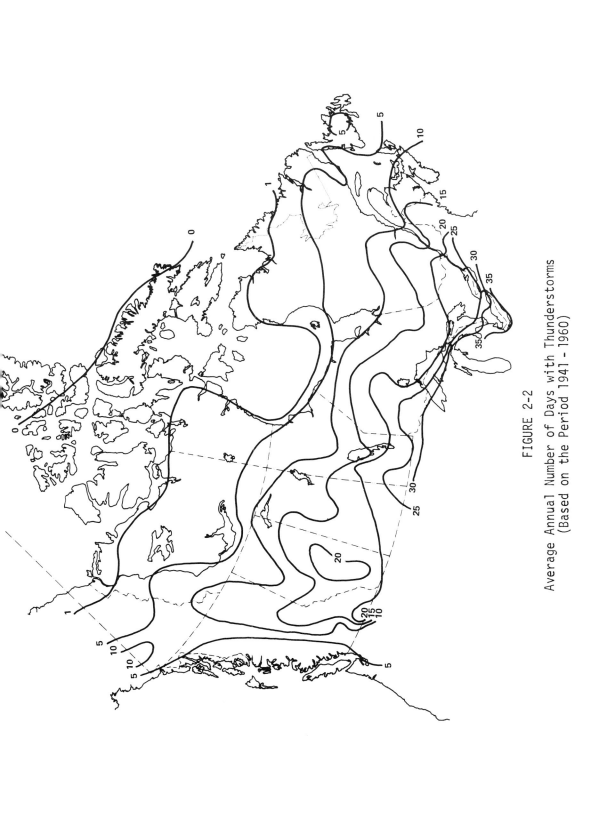

FIGURE 2-2

Average Annual Number of Days with Thunderstorms
(Based on the Period 1941 – 1960)

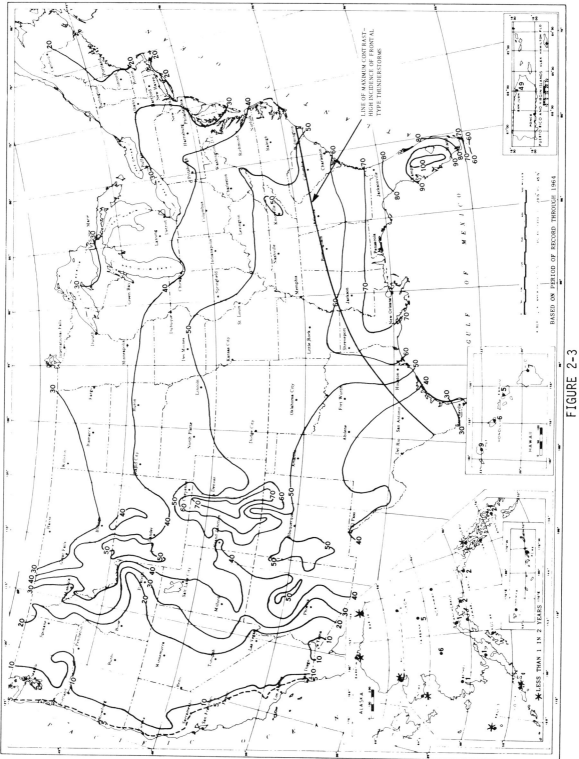

FIGURE 2-3

2.1.5 Mechanism of the Lightning Stroke

Photographic observations indicate that a lightning discharge may be initiated either from a cloud or from the peak of a tall structure on the ground[4]. The polarity of the lightning discharge or stroke is of little practical concern to the protection engineer, because a surge propagated through an object in either direction has the same effect. As a matter of interest, however, studies indicate that while strokes to ground occur from either positively or negatively charged clouds, the great majority of strokes occur between negatively charged clouds and corresponding positive charges on the earth. The electron flow of current is therefore from cloud to earth.

The lightning discharge is usually initiated from the charge center in the cloud in ionization steps called a *stepped leader*. In the case of discharges involving tall structures, the leader may be initiated at the structure. If it is assumed that a negative cloud is discharging to positive earth, the stepped leader path is pre-ionized by a steadily moving *pilot leader*, which leaves no visible track. The stepped leader follows this pilot leader path in approximately 50-yard increments at 50 μs intervals, with accompanying luminosity at each step as illustrated in Figure 2-4. The illustration is similar in appearance to the discharge traces that have been recorded on fast-moving film.

When the stepped leader reaches the earth, the neutralization of the negatively charged channel begins at the earth and travels in a brightly luminous path or *main stroke* toward the cloud at 10 percent of the velocity of light. The upward progress of the main stroke toward the cloud is accompanied by a rising surge of electron current from cloud to earth. The main surge of current, which lasts for less than a millisecond, may be followed by a low current lasting for approximately 0.1 second. There may then be a second leader, called a *dart leader*, which originates in a different part of the cloud charge center and follows the same path to earth but does not exhibit the stepped character of the first leader. The dart leader may result in a subsequent stroke; this may be followed by a second dart leader and so on. The average number of strokes in multiple discharges is about four — single-stroke discharges are quite common, and discharges having more than six strokes are rare. The median number of strokes per multiple discharge is two[5]. The time interval between strokes in a multiple discharge is so short (0.02 second) that several strokes may appear to the eye as one bright stroke. For design tests of the 'lightning withstand capability' of plant items and associated equipment, both in the communication and power industries, a single large impulse is employed. This is an 'equivalency' type of test dictated by practical test considerations. Experience has indicated, however, that this is an acceptable simulation of actual field exposure, which includes multiple component strokes.

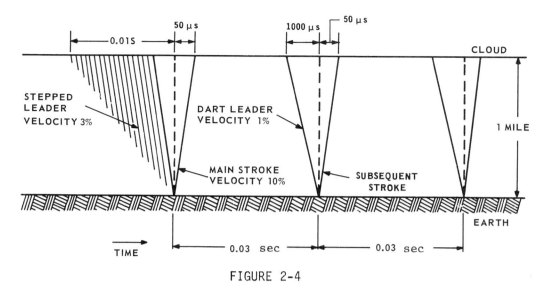

FIGURE 2-4

Mechanism of Lightning Strokes with Approximate Time Relations,
and Leader Velocities with Respect to the Speed of Light

2.1.6 Lightning Magnitudes and Waveshapes

Potentials associated with a lightning discharge are of very high magnitude. Estimates of potentials required to initiate discharge are in the order of 5 to 20 million volts. Were it not for the nonuniformity of the electrostatic field between cloud and earth, potentials exceeding these values would be required for discharge. The ionization of the charge center in the cloud initiates the pilot leader, which carries a charge with it toward earth. In this manner, the field at the tip of the leader becomes progressively more intense and usually proceeds until it contacts the earth or some object on it. Occasionally, however, the charge fed from the cloud to the leader is not sufficient to maintain its progress, and the charge dissipates without producing a complete stroke.

It is not the potential associated with a lightning stroke that concerns the protection engineer, however, but rather the damage to material objects on the earth from high-magnitude currents. Crest-current magnitudes will vary greatly from stroke to stroke, depending upon meteorological factors and the overall impedance of the stroke. Figures 2-5 and 2-6 are representative curves showing the distribution of crest magnitudes of stroke currents to aerial and buried plant.

The impedance of objects such as telephone plant, which may inadvertently become part of the path along which the electrical charges flow, is *considerably lower than that of the total path impedance*. Because of this fact, it is admissible and convenient to consider a stroke to be essentially a *constant current source* for any given stroke. Studies[5] indicate that the stroke channel impedance from cloud to earth is in the order of 5000 ohms, while resistance to ground of a point contacted by the stroke is relatively low, as defined by the following equation:

$$R_0 = \left(\frac{\rho E_0}{4J}\right)^{1/2}$$

where

R_0 = resistance to ground of point electrode

ρ = ground resistivity in ohms-meter

E_0 = surface breakdown gradient in kilovolts/meter

J = crest current in kiloamperes

Typical values in this equation for earth resistivity (ρ) ranging from a low of 100 ohms-metre to a high of 5000 ohms-metre and values of crest current (J) ranging from 8500 amperes to 100 000 amperes will only change R_0 from approximately 16 ohms to 121 ohms. When this low earth path resistance is compared to the aerial lightning channel resistance of 5000 ohms, it is apparent why the lightning mechanism and stroke channel appear as a relatively high impedance generator whose output is essentially 'constant'.

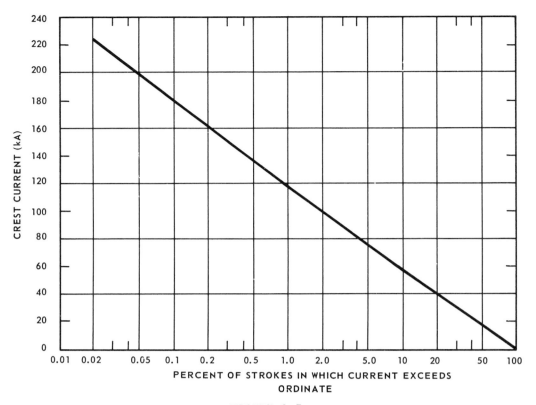

FIGURE 2-5

Distribution of Lightning Stroke Crest Currents
to Aerial Structures

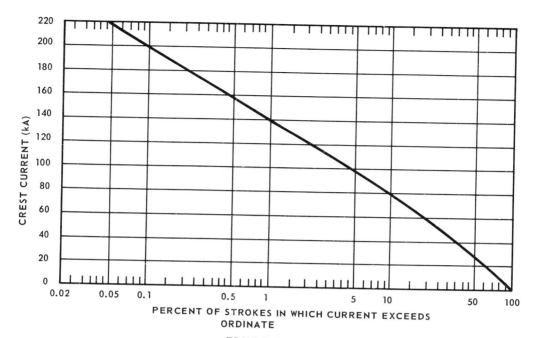

FIGURE 2-6

Distribution of Lightning Stroke Crest Currents
to Buried Metallic Objects

Measurements show that there are wide variations in rise and decay times between lightning strokes. Some strokes are characterized by rapid rates of rise, high crest currents, and relatively short decay times. Strokes of this type, sometimes referred to as 'cold' lightning strokes, leave no trace of burning or fusing but do develop very disruptive explosive pressures in materials such as wood poles where the vapor pressure developed cannot be readily vented. Strokes referred to as 'hot' lightning strokes generally have much smaller crest values but slower rise and decay times. These long-duration surges do not produce significant explosive effects but will ignite combustible material and fuse conductors. It is customary to define the waveshape of a lightning surge by two numbers such as 5×20 (five by twenty) where both values express time in microseconds (see Figure 2-7). The number 5 represents the rise time from zero to crest value, and the number 20 represents the subsequent surge decay time interval from zero to 50 percent of the peak surge value on the wavetail.

The average charge lowered in a stroke to open ground is estimated to range from 10 to 50 coulombs. In the Empire State studies[8], charges as high as 164 coulombs were recorded, but in studies of strokes to structures of lower height[9] the maximum charges recorded did not exceed 60 coulombs. *It would appear, therefore, that the charge on the structure contacted by the stroke increases with the height of the structure.*

The following quantitative information concerning lightning strokes[10] was summarized from data secured by several investigators through measurement of strokes to transmission towers and indicates order of magnitudes.

Median number of strokes in multiple discharges	2
Median time interval between strokes in multiple discharge	0.02 second
Median crest current	16 kA
Maximum crest current	220 kA
Median rate of rise of current	10 kA/μs
Median time for current to drop to half of its crest value	43 μs
Median total charge in stroke	30 coulombs
Maximum total charge in stroke	164 coulombs

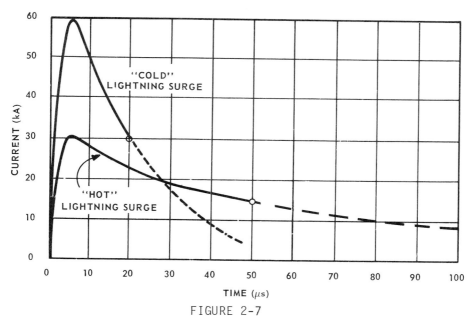

FIGURE 2-7

Representative Lightning Current Surges

2.1.7 Soil Resistivity

The resistivity of soil is a major parameter in protection problems such as those listed below, which involve earth conduction effects:

   a) Grounding problems — estimating the resistance to earth of grounding electrodes; prospecting for suitable locations for the construction of grounding structures (resistance of a grounding electrode varies directly with soil resistivity).

   b) Problems connected with earth return circuits, mutual impedance, shielding, etc.

   c) Buried cable problems — distance that remote strokes to earth will develop an arc path to a cable (arcing distances to a cable and also voltage induced in the core conductors of a cable vary as the square root of soil resistivity).

Soil, electrically, is of the nature of a nonlinear resistance that tends to decrease with increasing current densities. Soils consist of particles that may differ greatly in composition as well as in moisture and chemical content. The surface and subsurface conditions vary to a considerable extent; consequently, a wide range of resistivity values is possible even for soils that appear to be of the same general type. Because of this wide variation in the types and contents of soils constituting the earth's crust, it should be immediately recognized that we are dealing with a very nonhomogeneous medium, especially over different areas and depths.

When approximations are acceptable, it is common practice to determine by inspection the general nature of the surface and sub-surface soil types. A convenient way to secure data on the subsurface soil conditions is by inspecting excavations or cuts in the area under consideration. With this approximate information, some rather loose assumptions concerning resistivity may be made by means of generalized data such as those given in Table 2-1. Another method of estimating an effective resistivity value for use in engineering formulas is by means of geological maps, which show variations in the general structural nature of the earth's crust. Field measurements have indicated a reasonably consistent relationship between resistivity and the physical formations of the earth[18]. Unfortunately, geological maps generalize conditions over rather large areas and may not be particularly accurate with respect to a small area, but for problems involving comparisons of large areas, for example, a study of probable lightning trouble rates for certain types of equipment in Eastern Canada as compared to Western Canada, geological maps would provide a satisfactory basis for determining resistivity values. Such maps for the North American Continent may be secured from the Geological Society of America. Similar maps are probably available for other developed areas. Figures 2-8 and 2-9 give conductivity maps of Canada and Southern Ontario respectively.

TABLE 2-1

Range of Resistivity Values for Several Types of Soils

PHYSICAL COMPOSITION	RESISTIVITY IN OHMS-METRE
Sea Water (Reference)	1-2
Marsh	2-3
Fresh Water Lakes	70-125
Clay	3-160
Clay mixed with sand and gravel	10-1350
Chalk	60-400
Shale	100-500
Sand	90-800
Sand and gravel	300-5000
Rock (normal crystalline)	500-10000

NOTE: In engineering formulas, earth resistivity (ρ) is generally expressed in ohms-metre which is the resistance between opposite faces of a one-metre cube of soil.

The values given in Table 2-1 were taken from the appendix of a book by British authors and are presumably based on measurements of soils in their area. The data given agree closely with that of investigations carried on in the United States and Canada.

The resistivity of a given type of soil will vary materially with the moisture content, the concentration of dissolved salts, and soil temperature. When the moisture content of soil drops below about 25 percent by weight, the resistivity increases rather rapidly with further decreases in moisture content[17].

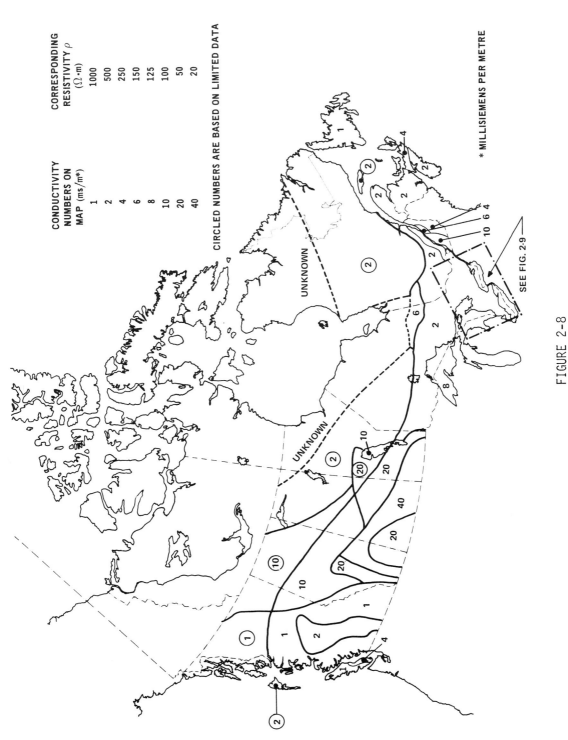

FIGURE 2-8

Soil Conductivity Map of Canada

CONDUCTIVITY NUMBERS ON MAP (ms/m*)	CORRESPONDING RESISTIVITY ρ ($\Omega \cdot$m)
1	1000
2	500
4	250
6	150
8	125
10	100
20	50
40	20

CIRCLED NUMBERS ARE BASED ON LIMITED DATA

* MILLISIEMENS PER METRE

SEE FIG. 2-9

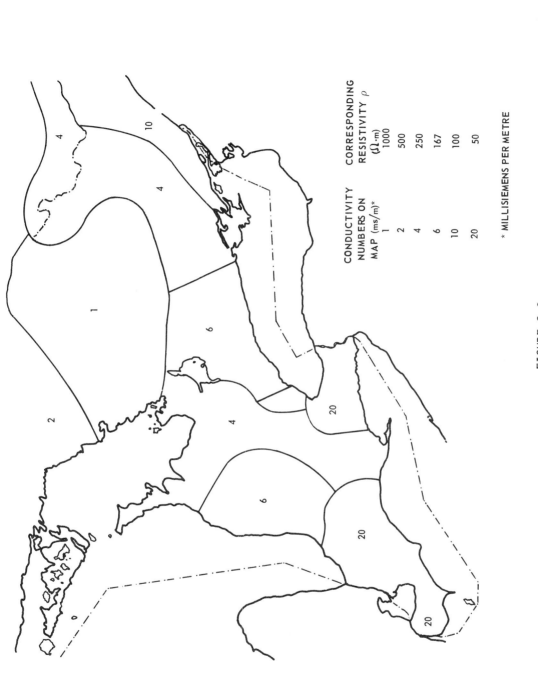

CONDUCTIVITY NUMBERS ON MAP (ms/m)*	CORRESPONDING RESISTIVITY ρ (Ω·m)
1	1000
2	500
4	250
6	167
10	100
20	50

* MILLISIEMENS PER METRE

FIGURE 2-9

Soil Conductivity Map of Southern Ontario

Variations in the moisture content of soils during seasonal cycles will change the surface resistivity and thereby influence the performance of a grounding structure having a significant portion of its structure in this surface layer. However, in situations where soil resistivity is the controlling factor to considerable depths, such as in the case of coupling problems, changes in surface moisture content may be neglected. Soil resistivity is also sensitive to temperature changes, especially in the transition period when moisture changes from liquid to ice.

2.1.8 Ionization Producing Soil Breakdown

When a volume of earth becomes energized, potential gradients appear both on the surface and in the soil itself. For example, gradients appear in the soil at the point where a lightning stroke contacts the earth, and generally will have sufficient magnitude to produce arcing in and on the surface of the earth around the point of contact. Evidence of this has been observed in the form of fused masses of soil referred to as Fulgurites. A potential gradient between a stroke point and a buried conductor some feet away will frequently produce an ionized path between the two, through which arcing takes place. Ionization breakdown can also occur in the soil adjacent to a ground rod where the potential gradient becomes extremely high due to concentration of current in the limited volume of soil close to the rod.

Soil ionization is a related factor in some types of protection problems for reasons such as those listed below:

1) An ionized path has a much lower resistance than an ordinary conduction path.

2) A grounding electrode, such as a rod, will have a significantly lower resistance to earth if ionization occurs that, in effect, increases the diameter of the rod and thereby the area in contact with the soil.

3) An ionized path between two objects in the earth permits currents to concentrate on small surface areas at the points of contact, thereby producing excessive localized heating that frequently results in fusing and the buildup of explosive pressure.

Measurements show that surface ionization occurs at fields ranging from 30 kV/ft to 130 kV/ft, depending upon local conditions. Considerably higher values of field are required, however, to produce internal ionization of soil.

The critical impulse breakdown value of soils is chiefly a function of physical composition. Soil breakdown measurements made in connection with Bell System cable protection studies, employing a moist mixture (28 percent H_2O) of 75 percent clay and 25 percent fine sand and a 1-1/2 × 40 microsecond impulse wave-shape, produced breakdown values in the order of 650 kV/ft. These values apply when breakdown occurs on the peak of the wave; however, breakdown may occur at somewhat lower potentials with slower wave fronts or after the voltage has passed its peak, although the ionizing time may be appreciable. For example, with moist clay the ionizing time may be in the order of 8 microseconds or more.

TABLE 2-2

Critical Impulse Breakdown Gradient of Soils[18]
(1-1/2 × 40 Microsecond Wave)

SOIL TYPE AND COMMENTS	IMPULSE (kV/ft)
GRAVEL, MOIST	
Breakdown on Tail of Wave	350-380
Breakdown on Crest of Wave	540-590
GRAVEL, DRY	
Breakdown on Crest or not at all	630-700
SAND, MOIST	
Breakdown on Tail of Wave	400-490
Breakdown on Crest of Wave	520-700
SAND, DRY	
Breakdown on Crest of Wave or not at all	520-580
CLAY, PLASTIC	
Wave departed from standard; time to breakdown erratic from 1 to 8 microseconds	580-1180

It is apparent that rather broad assumptions must be made in selecting values of ionizing gradients in connection with protection problems. Measurements are time consuming and, because of the heterogeneous nature of most soils, the final results may still not be particularly accurate. In protection problems involving buried cables, buried shield wires, etc., the following approximations may be taken as representative of the denser types of soil:

> Surface ionization —   90 kV/ft
>
> Internal ionization —   550 kV/ft

For the lighter types of soil, values in the order of the following are satisfactory:

> Surface ionization —   70 kV/ft
>
> Internal ionization —   400 kV/ft

2.1.9 Plant Damage

Open wire lines are relatively free from lightning damage. However, a direct stroke having a long decay time may fuse open a line conductor at the arc contact point. Also, lightning sometimes strikes wood poles and causes severe shattering. This explosive effect is not uncommon when lightning stroke current propagates through a low conductivity material.

Cable faults produced by lightning can be of several types, depending on the makeup of the cable structure and the type of plant construction. Table 2-3 summarizes the kind of cable faults commonly experienced in various types of cable plant.

TABLE 2-3
Typical Lightning Faults in Cable

TYPE OF CABLE	TYPE OF CONSTRUCTION	SERVICE	TYPE OF DAMAGE
Paired (Conventional design)	Aerial Separate Right of Way	Trunk	1) Pin holes in external Polyethylene jacket 2) Core-shield dielectric breakdown (pair to pair failure less likely)
Paired (Conventional design)	Aerial Joint-use with power	Distribution	1) Pair-to-pair dielectric breakdown (core-shield failure less likely)

TABLE 2-3 (Cont'd)

TYPE OF CABLE	TYPE OF CONSTRUCTION	SERVICE	TYPE OF DAMAGE
Paired (Conventional design)	Buried	Trunk	1) Pinholes in external Polyethylene jacket 2) Core-shield dielectric breakdown (pair-to-pair failure less likely) 3) Moisture gains access to core of alpeth type cables
Paired (Conventional design)	Buried	Distribution	1) Pinholes in external polyethylene jacket 2) Pair-to-pair dielectric breakdown (core-shield failure less likely) 3) Moisture gains access to core of alpeth type cables
Coaxial (Filled Type)	Buried	Trunk and Distribution	High degree of immunity from adverse effects of moisture. Behaviour similar in other respects to conventional PIC cable
Coaxial (Air-dielectric Types)	Buried	Trunk	1) Pinholes in external polyethylene jacket 2) Pair complement exposed to dielectric breakdown similar to paired type cables 3) Coaxial tubes unlikely to be damaged by sparkover between inner and outer conductors, but air dielectric tubes are susceptible to crushing
Paired	Underground	Trunk and Distribution	Generally used in unexposed locations (urban areas). When installed in exposed locations, it can experience failures similar to buried plant
Coaxial	Underground	Trunk and Distribution	Same as for Paired Underground

TABLE 2-3 (Cont'd)

TYPE OF CABLE	TYPE OF CONSTRUCTION	SERVICE	TYPE OF DAMAGE
Paired	Submarine	Trunk and Distribution	1) Cable submerged in several feet of water will not be exposed to direct strokes 2) Exposure occurs at and near shore line where effects are similar to buried plant
Coaxial	Submarine	Trunk and Distribution	Same as above

2.1.10 Exposure Classification

Exposure of plant to lightning is a function of physical location within a specific area as well as of isokeraunic levels. Classification by location based on relative degree of exposure is as follows.

a) Metropolitan Environment (Urban)

Exposure very low to negligible. Buildings intercept lightning strokes. Extensive metallic piping systems disperse currents in earth and provide shielding. Multiplicity of cable shields bonded at each manhole in underground (UG) duct runs distribute extraneous earth currents.

b) Suburban Environment

Exposure from low to intermediate levels. Varying degrees of shielding provided by buildings and structures, which in some cases may approach that of a metropolitan area. Equipment associated with outside plant consisting of a minimum of three metallic shielded cables in a common duct run is generally considered to have low exposure (essentially that of a metropolitan area).

c) Rural Environment

Usually high exposure, depending on incidence of thunderstorms, relative elevation of terrain, and soil resistivity. Shielding is minimal, so communication plant is subject to the full effects of lightning stroke currents.

2.2   EVALUATION OF LIGHTNING EXPOSURE CONDITIONS

2.2.1   Stroke Incidence to Plant and Structures

With respect to the diversion of lightning strokes to objects, the results reported by several investigators[6] vary considerably. This is understandable when working with a very limited sample in a highly variable environment. However, taken together the information has considerable engineering value.

It appears from a correlation of such data that the distance over which a sizeable proportion of the strokes are likely to be diverted to an aerial line ranges from about 3 to 5 times the height of the line. This ratio is higher for strokes of negative polarity and tends to vary inversely with the height of the line. A ratio of 3.5 is frequently mentioned as applicable to power transmission lines having heights from about 60 to 100 feet, and it would appear that a value approaching 5 is suitable for relatively low conductors such as an aerial communications cable having a height of about 18 feet.

In power literature, a 'rule of thumb' value for *direct strokes to transmission lines* of 1 stroke/mile/year is frequently mentioned for areas having an annual thunderstorm day incidence of 30 to 40. This approximation is probably based on S.K. Waldorf's observations[12,13] and appears to be an acceptable approximation as indicated by the following example.

Annual stroke incidence per mile of line,

$$P_a = F_s(TD) \ \frac{2(F_d \times h)}{5280} \ \ell$$

$$= 0.3 \ (35) \ \frac{2(3.5 \times 80)}{5280} \times 1 \qquad\qquad (2.1)$$

$$= 1.1 \ \text{strokes/mile/year},$$

where

F_s = Stroke factor[2] = 0.3

TD = Annual incidence of thunderstorm days = 35

F_d = Stroke diversion factor = 3.5

 h = Height of line = 80 feet

 ℓ = Unit length of line under consideration = 1 mile

2.2.2 Aerial Cables

In a similar manner using expression (2.1), the annual stroke incidence to a directly exposed aerial cable may be approximated.

Assumptions:

F_s = 0.28

TD = 15

F_d = 4.5

 h = 18 feet

 ℓ = 100 miles

$$P_a \simeq 0.28 \ (15) \ \frac{2(4.5 \times 18)}{5280} \times 100$$

$$\simeq 13 \text{ strokes/100 miles/year}$$

Sunde[5] (p. 323) states that aerial cable routes are shielded to a considerable extent by the presence of trees, which usually extend above the cable and divert strokes, so that the incidence of direct strokes to the cables may be smaller than would otherwise be expected. In addition to possible shielding from trees, joint-use aerial plant enjoys substantial shielding from the overbuilt power conductors. Where specified plant clearances are not observed, however, heavy lightning strokes may flash over to the telephone plant and establish a path for power follow current. The common practice of bonding telephone cable shields to power neutrals should mitigate the possibility of flashover to telephone facilities.

Exposure conditions are not uniform along a cable route as assumed in theoretical estimates, but averaging over a 100 mile length tends to compensate for variations. Also, where some knowledge of field conditions along a route is available, intro-duction of a judgment factor (terrain factor) should improve exposure estimates. In joint-use, therefore, the probable incidence of direct strokes obtained in formula 2.1 (predicated on direct exposure) will be considerably on the high side. It should be evident that such probability figures are only rough approximations; nevertheless, estimates so derived are useful in the process of determining levels of protection that can be economically justified.

2.2.3 Structures

The incidence of *strokes to structures* (such as a radio antenna support tower) can be estimated as in aerial cables. The critical circular area from which strokes are diverted to an antenna mast in the range of heights frequently encountered in practice (100 to 500 feet) is roughly a circle having a radius of 3 to 4 times the effective height of the structure. On flat terrain, the effective height is that of the tower, but for a similar structure on a prominent hill the effective height would approach that of the tower plus the height of the hill. The following example will illustrate the method of estimating the probable annual incidence of strokes to a 150-foot tower on a hill having an elevation of 200 feet in an area having an isokeraunic level of 30.

$$P_b \simeq F_s \ (TD) \ \frac{\pi(F_d \times h)^2}{27.9 \times 10^6}$$

$$\simeq 28 \ (30) \ \frac{3.14(3 \times 350)^2}{27.9 \times 10^6} \tag{2.2}$$

$$\simeq 1.1 \ \text{strokes/year}$$

Curves of estimated annual stroke incidence versus height of mast are given in Figure 2-10.

2.2.4 Buried Cables

Lightning strokes may initially contact the earth and then side flash to well-grounded buried objects, especially in high resistivity soil. E.D. Sunde,[5] p. 298, gives the following formulas for approximating *critical arcing distances to buried objects:*

$$\frac{\rho \leq 100 \ \text{ohms-metre}}{r = 0.26 \ (J\rho)^{\frac{1}{2}}} \tag{2.3}$$

$$\frac{\rho \leq 1000 \ \text{ohms-metre}}{r = 0.15 \ (J\rho)^{\frac{1}{2}}} \tag{2.4}$$

where

    r = critical arcing distance in feet

    J = stroke current in kiloamperes

    ρ = soil resistivity in ohms-metre

Typical arcing distances for a 30-kiloampere stroke (average stroke to buried structures,[5] p. 294) are given in Table 2-4.

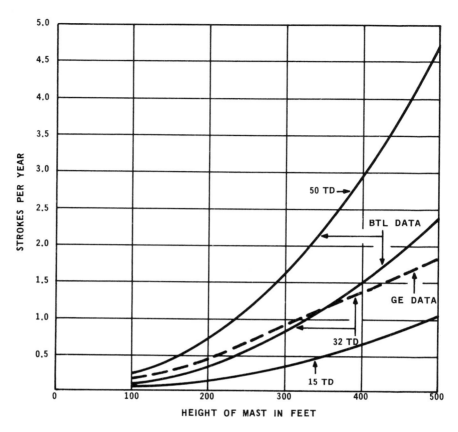

FIGURE 2-10

Incidence of Strokes to a Radio Mast

TABLE 2-4
Arcing Distance vs Soil Resistivity

SOIL RESISTIVITY (ρ) IN OHMS—METRE	ARCING DISTANCE IN FEET
100	14
250	17
500	20
1000	26

In Table 2-4, homogeneous soil conditions and absence of grounded objects other than the one under consideration have been assumed. In the case of a buried communication cable, actual field conditions may be quite different due to the presence of foreign objects such as metal fences and trees (especially the former), which are known to substantially increase the exposure. Also, in hilly terrain, a cable routed along a ridge may have comparatively high exposure. When some physical details such as these are known, the introduction of a 'terrain factor' may improve the accuracy of the exposure estimate. This would be a matter of judgement based on the apparent degree that the exposure conditions deviate from the average or unity. An estimate is given by

$$P_c = K_T (F_s) \ (TD) \ \frac{2(d)}{5280} \ \ell \qquad (2.5)$$

where

K_T = Terrain Factor*

F_s = Stroke Factor

TD = Annual incidence of thunderstorm days

d = Critical arcing distance in feet

ℓ = Unit length of cable = 100 miles

The following is an example of nominal stroke incidence to a buried cable for a typical situation:

$K_T = 1$

$F_s = 0.28$

$TD = 15$

*For flat, uniform terrain without the presence of perturbing objects (nominal condition), the terrain factor would be unity.

d = 17 ft (see Table 2-4, ρ = 250 ohm-metres)

ℓ = 100 miles

Using expression (2.5),

$$P_c = 1(0.28)\ (15)\ \frac{2(17)}{5280} \times 100$$

$$= 2.7 \text{ strokes per year per 100 miles of cable.}$$

The incidence of strokes to *underground (UG) cables* will be essentially the same as that to buried cables where exposure conditions are similar. Trouble records confirm that UG cable runs in rural areas are subject to direct lightning strokes that may explode the duct in the process of reaching the cable shield. However, lightning troubles in the overall UG plant are small because much of this plant is in urban areas.

2.2.5 Submarine Cables

Submarine cables are in general less subject to direct lightning strokes except at the shore line. Excluding pure water, water generally, and especially that high in contaminants, has a relatively low resistivity. If a resistivity of 70 ohms-metre is assumed for lake or river water, the critical depth below which lightning is unlikely to arc directly to a submerged conductor is given in Reference 5, p. 299, as

$$\text{Arcing depth } d \simeq 0.3\,r \text{ feet} \tag{2.6}$$

where

$$r = 0.26(J\rho)^{\frac{1}{2}} \text{ (see equation 2.3).}$$

For a stroke of 100 kA and ρ = 70, the penetration of a stroke to the cable would be

$$d \simeq 0.3(21.7) \simeq 6.5 \text{ feet.}$$

2.2.6 Correlation Between Theoretical Predictions and Reported Plant Troubles

Development of a model for predicting cable troubles in aerial exchange plant presents many practical problems due to the large number of physical variables involved. Buried plant is more predictable, and consequently published analytical work is chiefly restricted to this type of plant. Sunde recognized this problem, and

therefore chose to explore the surge behavior of exchange plant by
experimental studies on a test line. This led to the so-called
'New Brunswick (New Jersey) Studies'. Data on electrical troubles
experienced in the operating cable plant may be obtained from Bell
Canada, "Cable Summary Data". Data abstracted from these
forms for the year 1971 and a breakdown of troubles (lightning
versus power contacts) obtained from supplemental reports from some
Bell Canada areas are given in Tables 2-5 and 2-6 respectively.
These data give some insight into the actual fault situation and
provide a basis for evaluating the accuracy of theoretical estimates
based on stroke incidence predictions obtained by procedures outlined
earlier in this subsection. With regard to trouble report forms, it
is established practice to report each sheath opening requiring
repairs as a 'case of trouble'. For example, a lightning stroke or
power contact may produce multiple troubles, in some cases only a
few feet apart, yet each sheath opening requiring repairs is
reported as a case of trouble.

The degree of correlation between derived stroke incidence
figures and actual trouble experience has been explored. Using a
set of conditions judged to be roughly representative of the major
operating areas of Bell Canada, the following stroke incidence
figures were obtained (refer to Aerial Cables and Buried Cables in
this subsection).

Aerial Cable — 13 strokes/year/100 miles of cable

Buried Cable — 2.7 strokes/year/100 miles of cable

Lightning cable troubles reported by Bell Canada for 1971 are
given in Table 2-5. Annual trouble rates per 100 sheath miles are
5.2 for aerial and 1.7 for buried.

It is evident that theoretical stroke incidence to *unshielded*
aerial plant does not agree with actual experience. Some obvious
reasons for this variance are:

1) Major shielding from joint-use construction, and possibly
     to a lesser degree some shielding from trees, structures,
etc.

2) Lightning currents on cable shield are restricted to
     shorter sections due to shield grounding practices such
as bonds to power neutrals, and grounded guys.

3) A sizeable portion of the plant is now of the PIC
     (polyethylene insulated conductor) type, which withstands
substantial voltages. Consequently, stroke currents at the
lower end of the magnitude distribution (Figure 2-5) do not
produce critical voltages.

In the case of buried plant, the correlation between theoretical stroke incidence and reported trouble rate is substantially better, probably because we are now dealing with a more predictable set of physical conditions. Some apparent factors that may account for the reported trouble rate being about 40 percent less than estimated stroke incidence are:

1) In built-up areas, the plant may benefit from some fortuitous shielding.

2) Buried plant is largely of the PIC type, which withstands some of the smaller stroke currents.

It is generally believed that contacts between energized power conductors and telephone plant are a minor factor in overall electrical cable faults. There is not much data to support this opinion, which in itself may be significant. Limited surveys have been made, resulting in data such as that given in Table 2-6. A comparison of troubles resulting from power contacts versus lightning troubles using data from Table 2-6 is given in Table 2-7. It appears that power-produced faults represent roughly 6 percent of all reported electrical troubles. Relief from power contacts is chiefly obtained by high reliability power line construction and physical isolation of the telephone plant as much as practicable.

When making estimates of cable trouble rates from generalized information, it is obvious that the physical variables are difficult to evaluate with a fair degree of accuracy, but it might be presumed that the dielectric strength of the cable facility can be predicted with good accuracy. Such may be the case of cable on a reel before installation; however, experience indicates that, due to handling and variations in splicing skill, an 'inplace' facility will have a substantially lower dielectric strength than would appear from a cable factory test specification. It should be evident, therefore, that generalized predictions of trouble rates can at best be only gross approximations.

TABLE 2-5

Summary of Lightning Troubles in
Bell Canada Plant — 1971

TYPE OF PLANT	SHEATH MILES	CASES OF ELECTRICAL DAMAGE	ELECTRICAL TROUBLES/YR /100 MILES
Aerial	54352	2806	5.2
Underground (duct)	10314	174	1.7
Buried	33469	568	1.7
TOTAL	98135	3548	3.6

TABLE 2-6

ELECTRICAL CABLE TROUBLES

Comparison of Lightning and Power Contact Troubles
(Limited Number of Areas Reporting)

AREAS	SHEATH MILES			CASES OF TROUBLE		TROUBLES/YR/100 MILES		POWER TROUBLES AS PERCENT OF TOTAL TROUBLES
	AERIAL	UNDER-GROUND (DUCT)	BURIED	LIGHTNING	POWER CONTACTS	LIGHTNING*	POWER† CONTACTS	
CENTRAL 1968	9466	884	4118	981	48	6.8	0.51	4.7
1969	10200	957	4234	927	33	6.0	0.32	3.4
1971	13446	1186	5984	585	24	2.8	0.18	4.0
AVERAGE	11037	1009	4779	831	35	5.0	0.32	4.0
WESTERN 1969	14014	2141	13100	951	70	3.2	0.50	6.9
TORONTO 1969	4983	2671	2175	116	16	1.2	0.32	12.1

*Lightning Trouble rates based on total sheath miles.

†Power Trouble rates based on aerial sheath miles.

TABLE 2-7

Power vs Lightning Troubles in Cable Plant

TYPE OF CONSTRUCTION	SHEATH MILES	LIGHTNING TROUBLES	POWER TROUBLES	TROUBLE/YR/ 100 MILES
Aerial	30034	–	121	0.4
UG	5821	–	–	–
Buried	20054	–	–	–
TOTAL	55909	1898	–	3.4*

* Lightning only.

Cable troubles reportedly caused by power may have resulted from direct contact between the cable and an energized power conductor, or from power-follow current in a path initially established between the power and telephone plant by a lightning flashover.

Based on the reported data summarized in Table 2-7, it appears that power-initiated troubles account for only about 6% of all reported electrical troubles.

2.2.7 Surge Parameters — General

The characterization of lightning surges in electric circuits and laboratory simulation for testing of components and facilities was initiated by the power industry. Much basic information concerning *lightning in aerial conductors* has been obtained by the power utility industry and manufacturers of transmission and distribution equipment. There are many AIEE and IEEE papers on the subject, and this published information is credibly abstracted in Reference 6. These investigations on power facilities were prompted by necessity, but apparently the same degree of urgency was not exhibited by the communication industry in America to obtain comparable exposure data, with the exception of disturbances in the noise range. Interest on the part of telephone companies in lightning exposure conditions has increased in more recent years because of the relatively high susceptibility to lightning damage of semiconductor components and miniaturized circuitry, but attention has been directed chiefly to cable plant.

The international power community (IEC) have standardized the waveshape of a voltage surge in overhead transmission facilities for test purposes as a 1.2 × 50 impulse.* The standard test wave in North America until a few years ago was 1.5 × 40, and experience indicates that power equipment so tested has performed satisfactorily in the field. Since a 1.2 × 50 wave is within the tolerances allowed under the old ASA standard, the change was made in American standards to bring them into agreement with the international standards.

The stroke path from cloud to earth is predominantly resistive (see 2.1), and surge impedance is in the nature of resistance; therefore, surge current should have essentially the same waveshape as the associated voltage surge. For industrial testing, practical considerations necessitate the use of surge current waves having slower fronts than obtained in practice. Test waves such as 8 × 20 are sometimes used for testing power equipment. To produce current wave fronts in the range of 1 to 2 microseconds would require a resistance many times larger than all the impedance of the test specimen to be placed in series with the generating source. This would reduce generator output efficiency so drastically that excessively large and costly surge generators would be required. Consequently, surge current test waves presently employed for industrial testing do not give direct simulation but rather a form of equivalency, the validity of which has been confirmed by satisfactory field performance.

Some of the information obtained by the power industry can be useful to the communications engineer, chiefly as background material, but one must recognize the major differences between the two types of facilities and the differences in exposure and protection devices employed. Some degree of similarity exists in the area of open wire lines, but in general the basic parameters are much different, and this must not be overlooked.

2.2.8 Cable Plant — General Considerations

Lightning continues to be the major cause of electrical troubles in telecommunication systems. The extensive use of plastic-insulated cable (PIC) has substantially reduced lightning produced cable troubles, but the primary exposure to lightning remains unchanged. PIC cable, because of its higher dielectric strength, can support higher amplitude surges, which tend to expose connected apparatus to even greater stress than was the case with paper-insulated cable.

The field investigations that produced the data to be discussed in the following sections were conducted on 'nonfilled' types of cable. However, it is reasonable to assume that the information obtained is also valid for 'filled' type cable, since its structure and electrical properties are not significantly different.

* See Glossary for definition.

E.D. Sunde's book[5] is a fundamental reference on the behavior and effects of lightning in communication plant. Although it is oriented primarily to buried cable, open wire and aerial cable are also covered. The theory is rigorous, and the numerical solutions are valid for the assumed conditions. However, *it is important to note* that the material in his book predates the field studies presented in the Trueblood – Sunde Paper[2]. These studies established that lightning surge currents in the shield of a buried cable are more likely to have decay times to half crest value in the range of 130 to 1000 microseconds rather than the 65 microseconds used by Sunde for the examples in his book. There will be further discussions of the Trueblood – Sunde paper in the section on Buried Cable.

The major sources of surges on the core conductors of paired cable are:

1) Induction from lightning surge currents flowing on cable
    shields. This is the predominant surge source in the
    case of trunk cables.

2) Surges from wire extensions.

3) Earth potentials produced by strokes in the vicinity of
    station grounds.

4) To a lesser extent, strokes to station drops.

2.2.9 Open Wire Communication Lines — Surge Parameters

GENERAL. Material published by investigators in South Africa[10a] is an excellent source of information on lightning in open wire communication lines. Their environmental conditions cover a relatively broad range in both isokeraunic levels and soil resistivities. More recent Canadian Studies[12] include some measurements on open wire lines, but not in the same depth as in South Africa, where open wire lines are still extensively used. Very recent data obtained by the Japanese[20] on nonmetallic sheath cable also indicates that such plant exhibits some of the characteristics of open wire facilities with respect to surge waveshapes.

SURGE VOLTAGE MAGNITUDES. The ceiling value of surge voltages that a facility can support is determined by the breakdown value of its insulation, which in the case of open wire lines is relatively high. The following statement from Reference 10a, page 16, is of interest with respect to the probable flashover level of a working line: "Impulse tests on standard telephone lines reveal that, in practice, flashover may occur at 30–40 kV and not at 70–80 (subscriber lines) and 110-120 kV (trunk lines) as mentioned above. This is due to the small clearances between the wires and the arms at points where J spindles are used for terminating conductors and also where stays are near the wire. Thus, although the traveling wave initiated by flashover may have a crest value of 70-120 kV, it will frequently be reduced by (subsequent) flashover to 30-40 kV by the time it reaches the end of a route." Now if the maximum initial flashover value (of 120 kV) quoted above is assumed, the maximum probable surge current in any one line conductor would be only 63 amperes. Sunde[5], page 337, suggests a ceiling flashover voltage of 500 kV for estimating surge current in an open wire line, in which case the surge current per conductor in a 10 wire line would be about 263 amperes. These are values of surge current that a protector unit may see at a point sufficiently removed from a stroke location where the line assumes a typical surge impedance.

The South Africans have not published a magnitude distribution of surge voltages. With respect to direct strokes, they merely state that line flashover will probably occur at about 40 kV with probably no damage to the line. They were principally interested in surge currents, and state that protective gaps that earth the line at voltages exceeding about 1000 volts are used, so current is of prime interest. Probable current magnitudes are discussed in a subsequent section of this text.

In Figure 3 of the Canadian studies[12], a "distribution of surge maximum peak voltage amplitude" is given, e.g., a value of 2000 volts is given for the 99.9 percent distribution point. This distribution is based on 6405 samples and because of the high sensitivity of the recording equipment, it seems likely that the sample does not include any significant number of direct strokes.

SURGE WAVESHAPES. Distributions of voltage rise and decay times in open wire plant are given in Reference 12 and significant values are presented in Table 2-8. The rise and decay times recorded in this study are substantially longer than generally assumed in the industry. Corresponding data from Reference 20 on nonmetallic sheath cable are also shown in Table 2-8. The rise and decay times obtained in this latter study are shorter and more in line with industry concepts. Reference 10a only gives waveshape information on surge currents. However, if it is assumed that voltage and current waveshapes are about the same, there is a fair degree of correlation between the values recommended in References 10a and 12 for the testing of apparatus and protection devices, as may be noted from Table 2-9.

TABLE 2-8

Waveshape Data Obtained on Open Wire Lines and Nonmetallic Sheath Paired Cable

SOURCE OF DATA	DISTRIBUTION POINT (%)	TIME TO CREST (μs)	DECAY TIME TO HALF CREST (μs)	RATE OF RISE ON FRONT OF WAVE V/μs
Canada[12] (Open Wire)	0.13	2.0		220*
	2.27	4.5		
	50.0	30.0	200	
	97.7		800	
	99.9		1000	
Japan[20] (Nonmetallic Sheath Cable)	0.13			670†
	2.27	1.2		
	50.0	5.0	18	
	97.8		70	

* This value has been derived on the assumption that the higher voltage surges were associated with the steeper wave fronts, i.e.,

approximate rate of rise = $\dfrac{1000 \text{ V}}{4.5} \simeq 220$ V/μs.

† On the basis of the same assumption,

approximate rate of rise = $\dfrac{800 \text{ V}}{1.2} \simeq 670$ V/μs.

TABLE 2-9

SOURCE	RECOMMENDED WAVESHAPE
Canada Study[12]	4 × 200
*South African Study[10a]	
Long Lines	10 × 800
Short Lines	2 × 100

* The investigators state that these values are
the result of strokes that did not directly
contact the line conductors. It must be concluded that the available data do not cover
all exposure possibilities.

SURGE CURRENT MAGNITUDES. In view of the lack of substantial data
regarding the probable ceiling magnitudes of surge current that may
flow to ground over an open wire conductor, either through a fault
or a protection gap, it is not surprising that there is considerable
divergence of opinion on the subject. Sunde[5] bases his estimate on
a 10-wire line having a surge impedance of 200 ohms and an assumed
flashover voltage between line conductors and ground of 500 kV, and
he obtained a figure of 2500 amperes. The South Africans suggest
flashover in the range of 40 to 120 kV, so on the same basis their
figures are 100 to 600 amperes. It would appear that, except in the
immediate vicinity of a direct stroke to a line, conductor surge
current is unlikely to exceed about 2500 amperes.

The South Africans suggest test values (Table 2-10) for
establishing the ability of a protector to provide a 5-year operating life on open wire facilities in an area having an isokeraunic
level of 70.

2.2.10 Aerial Exchange Cable — Surge Parameters

GENERAL. Analytical characterization of exchange plant is difficult because of the complex plant arrangements. For this reason,
in the late 1940's an extensive series of tests was conducted
on a 2-mile test section of aerial cable, which permitted surge
testing of a large number of exchange plant arrangements. This

TABLE 2-10

Surge Current Carrying Capability Tests for Protectors
Used on Open Wire Lines

PLANT	WAVESHAPE IN MICROSECONDS	CREST CURRENT (AMPERES)	NUMBER OF SURGES
Long O.W. Lines	10 × 800	50	50
		100	20
		300	5
Short O.W. Lines (less than 600 meters or 0.375 miles)	2 × 100	50	20
		500	5
		1000	2

Notes:

1. Assumed Conditions: $\rho = 1000$ ohms-metre   TD = 70/year.

2. Protector considered grounded when its resistance is less than 0.5×10^6 ohms at 200 volts (dc).

3. One direct stroke anticipated near a protector once in 5 years, in which case 'damage or short circuit may result. It is not generally economic to make telephone protectors which will withstand high current direct strokes without short-circuiting'.

4. The data for this table has been taken from Reference 10a.

5. The test values were based on a presumed 5-year life.

project was known as the 'New Brunswick Test', and the results were presented in a series of graphical charts showing the volt-current relationships in various representative arrangements. These tests were conducted prior to the introduction of PIC cable when it was not uncommon in high exposure areas for lightning troubles in paper insulated cables to occur at an annual rate of 50 or more per 100 miles.

This study is mentioned only for its possible academic value. Its practical value has been superseded by the extensive use of PIC cables, which are relatively self-protected. The survival of PIC cable is a relatively minor problem, so the major interest now is the exposure of connected apparatus. The common exposure modes of aerial exchange cable are illustrated in Figure 2-11.

SURGE VOLTAGE MAGNITUDES. Waveshape data obtained in field studies seems to have generated more interest than voltage magnitudes, probably because it is generally assumed that protectors will be used to place a ceiling on voltage amplitudes. Designers of apparatus to be used on paired cable believe that it is impractical to expose their devices to the exposure levels of the transmission facilities. A design ceiling is thus established, which is a value that adequately coordinates with the protectors to be used. Support for this approach rests on the fact that it requires several thousand volts to damage PIC cable and yet trouble reports (see Correlation Between Theoretical Predictions and Reported Plant Troubles) establish that such facilities are faulted by lightning.

The incidence of longitudinal surges decreases rapidly with magnitude. In the Canada Studies[12], surge voltages are equal to or greater than 1000 peak volts at the 99.9 percent distribution point. The four test cables (length 7 miles each) were located in areas having soil resistivities ranging from 10 to 150 ohms-metre and an average isokeraunic level of approximately 1.7. Assuming a soil resistivity value of 250 ohms-metre, which is representative of the Bell Canada operating area, these data indicate an expectancy of about three surges per 100 miles of cable per year, each surge being equal to or greater than 1400 peak volts.

SURGE CURRENT MAGNITUDES. Information on the longitudinal surge impedance of cable conductors and distribution of surge currents is meager. Some unpublished data tend to indicate that longitudinal impedances may range from roughly 100 to 300 ohms. Table 2-11 shows measured protector discharge currents. Details of test procedures are not known.

The 50-pair lead sheath cable is of more interest as it more nearly represents the size of cable in common use. In this case,

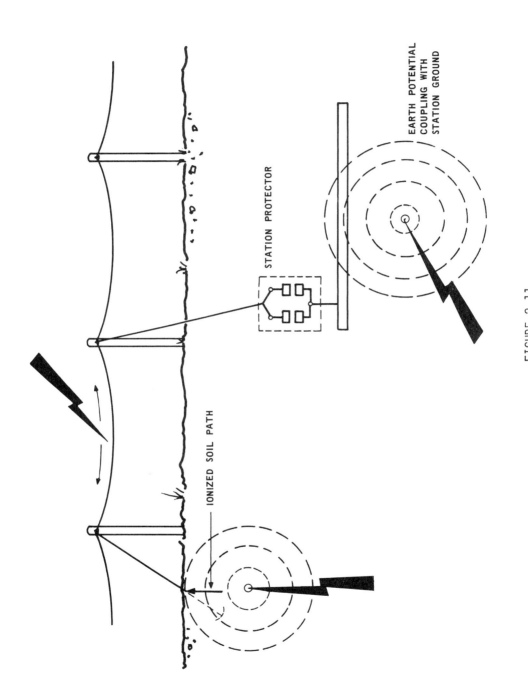

FIGURE 2-11

Exchange Cable — Common Modes of Exposure to Lightning

TABLE 2-11
Measured Protector Discharge Currents

TYPE OF CABLE	CURRENT RANGE (AMPERES)	NO. OF PROTECTOR DISCHARGES
Lead Shield (50 pr.)	10-25	464
	25-100	6
Polyjacketed Alpeth (10 pr.)	10-25	284
	25-100	27

only 1.3 percent of the discharges were outside the range of 50 to 100 amperes. It is estimated that this cable had a shield resistance of roughly 2 ohms/mile. Now, since core-shield voltage varies directly with shield resistance, it is reasonable to assume that, as a cable size approaches that of a full size cable (0.7 Ω/mile), discharge currents in the 50 to 100 ampere range would be substantially less than 1 percent. Analytical considerations offer some support for these figures. Cable insulation has limited dielectric strength (paper 1.5 to 2.0 kV and 'in place' PIC 5 to 10 kV) so, to support 100 amperes without breakdown, the longitudinal impedance would have to be somewhat less than normally assumed. It would appear that conductor currents in metallic shielded telephone cables are unlikely to exceed about 100 peak amperes.

Currents discharged through station protectors can be substantially larger than the values just discussed. (Refer to Summary of Surge Current Parameters.)

WAVESHAPE OF SURGES IN CORE CONDUCTORS. Mostly from deduction, the Bell Telephone Laboratories, USA (BTL) decided around 1955 to use a 10 × 600 impulse for protection testing. This waveshape was used for several years with apparent success, e.g., for the E6 repeater. In 1961, the results of field studies[11] designed specifically to characterize surge voltages in both aerial and buried paired cable suggested the substitution of a 10 × 1000 wave for test purposes. Since publication of that standard this somewhat longer wave has been generally adopted in North America as a standard wave for simulating longitudinal and metallic lightning surges in metallic shielded paired cable facilities. During the studies, 3-mil carbon block protectors were present on all pairs, including those monitored. This provided longitudinal waveshape information up to the operating voltage of protectors and, in addition, data was obtained on metallic voltages resulting from asymmetrical protector operation.

Subsequent investigations[12,20] have contributed additional information on this fundamental subject.

Data obtained in these three studies have been published in the form of probability distributions, which are most useful for engineering purposes. For the convenience of the reader, the original basic curves from these sources are given in the series of Figures 2-12 to 2-18. Further work has been directed to combining these data into single figures to facilitate comparisons (Figures 2-19, 2-20). Significant data points are summarized in Table 2-12.

Some general comments on the comparison of data are:

1) Wave front data for both aerial and buried cable check reasonably well from an engineering standpoint. Ten microseconds at the -3σ point appears to be a satisfactory generalization.

2) Decay times show major divergence. In the case of aerial cable, the correlation of the BTL and Japanese data is good. The Telephone Association of Canada (TAC) lightning data, however, show consistently longer decay times by factors of roughly 2.5 to 3. The cause of this is not evident and presents some uncertainty when deciding on a test wave for laboratory simulation.

3) In the BTL studies, there were relatively small differences between aerial and buried cable data. Statistically this difference is much greater in the TAC lightning survey.

The significance of these differences from an engineering standpoint is discussed in this subsection under the heading Parameters for Simulation Testing in connection with the selection of laboratory test waves.

2.2.11 Buried Trunk Cable — Surge Parameters

GENERAL. Burying a cable does not preclude its exposure to lightning, since buried cables offer an attractive 'sink' for strokes contacting the earth within a critical distance. The method of computing critical arcing distances to buried objects is covered in this subsection under the heading Buried Cables, and Figure 2-21 shows the manner in which a stroke to ground can ultimately contact the shield of a cable. Since surface ionization occurs at substantially lower voltages than breakdown through soil, arcing may occur over distances of a few feet up to about 100 feet.

In trunk plant, current flowing on the metallic shield is the predominant cause of longitudinal surges in the core conductors.

TABLE 2-12

Summary of Significant Data Points

AVERAGED DATA COMBINATIONS	TYPE OF CABLE PLANT	DISTRIBUTION POINTS	TIME TO CREST IN μs	DECAY TIME TO HALF OF CREST IN μs
BTL[11] & Japanese[20]	Aerial	-2σ	25	
		-1σ	50	235
		μ	98	400
		1σ		750
		2σ		1250
BTL[11], TAC[12] & Japanese[20]	Aerial	-2σ	25	
		-1σ	57	317
		μ	135	667
		1σ		1350
		2σ		2067
BTL[10] & BC[11]	Buried	-2σ	21	
		-1σ	33	350
		μ	78	685
		1σ		1390
		2σ		2300
BTL (for Comparison)	Buried	-2σ	21	
		-1σ	44	200
		μ	98	370
		1σ		780
		2σ		1200

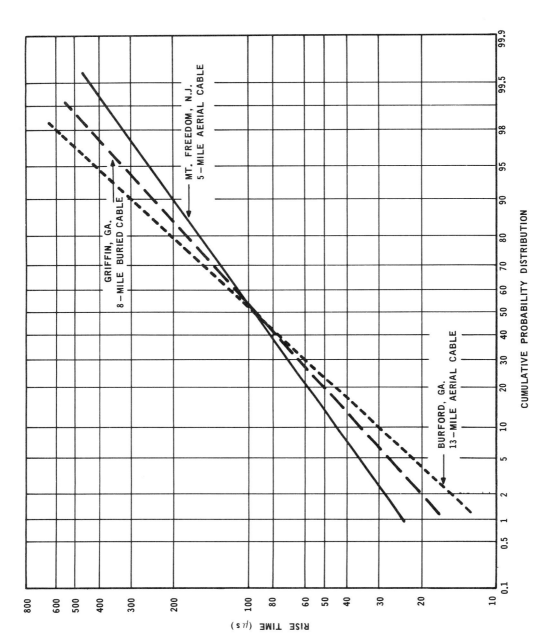

CUMULATIVE PROBABILITY DISTRIBUTION

FIGURE 2-12

Probability of Breakdown vs Rise Time

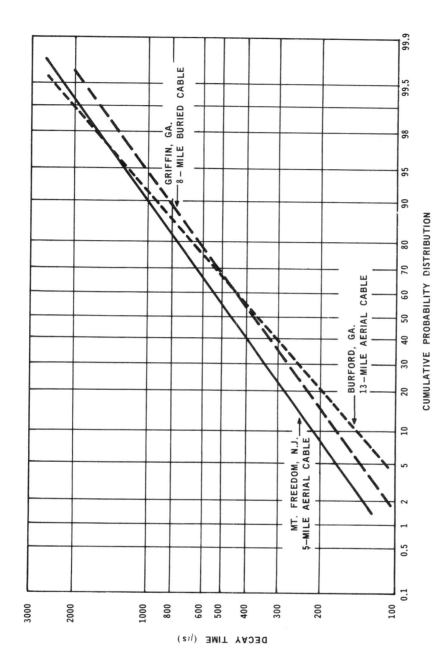

FIGURE 2-13

Probability of Breakdown vs Decay Time

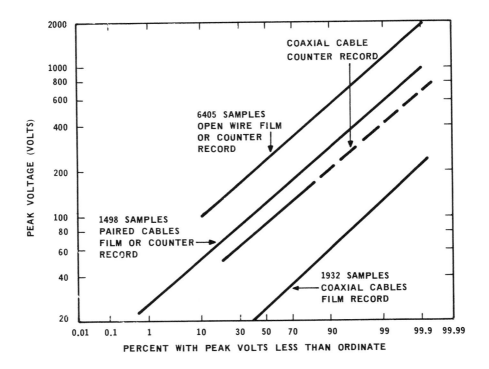

FIGURE 2-14

Breakdown Distribution vs Peak Voltage

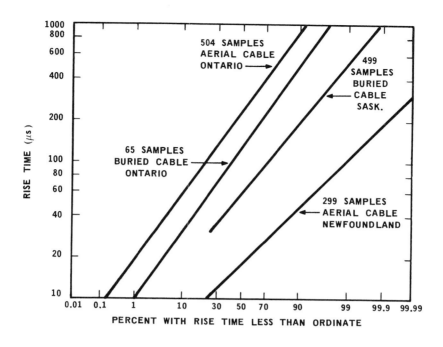

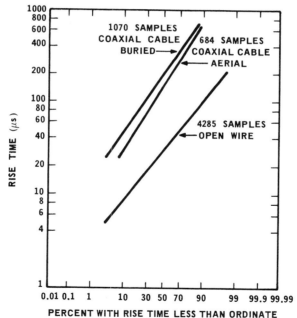

FIGURE 2-15

Breakdown Distributions vs Rise Time

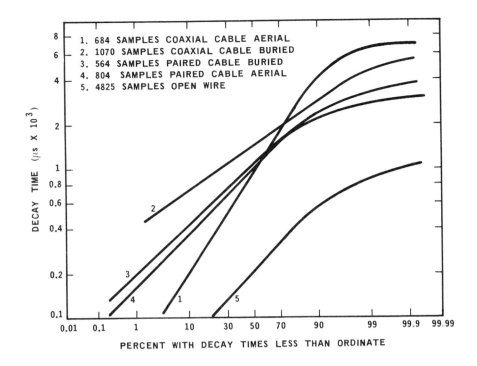

1. 684 SAMPLES COAXIAL CABLE AERIAL
2. 1070 SAMPLES COAXIAL CABLE BURIED
3. 564 SAMPLES PAIRED CABLE BURIED
4. 804 SAMPLES PAIRED CABLE AERIAL
5. 4825 SAMPLES OPEN WIRE

DECAY TIME (μs X 10^3)

PERCENT WITH DECAY TIMES LESS THAN ORDINATE

FIGURE 2-16

Distribution of Decay Times

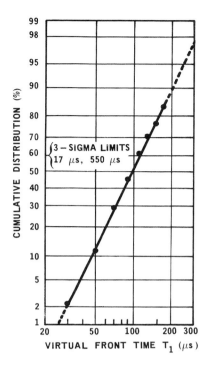

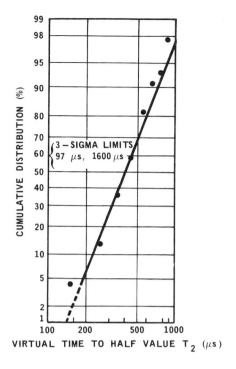

FIGURE 2-17

Distribution of Virtual
Front Times

FIGURE 2-18

Distribution of Virtual Times
to Half Value

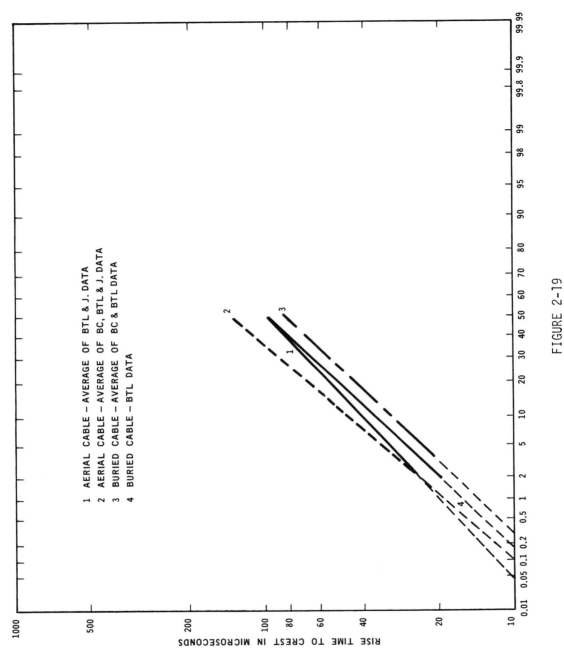

FIGURE 2-19

Distribution of Rise Times in Paired Telephone Cables

1 AERIAL CABLE – AVERAGE OF BTL & J. DATA
2 AERIAL CABLE – AVERAGE OF BC, BTL & J. DATA
3 BURIED CABLE – AVERAGE OF BC & BTL DATA
4 BURIED CABLE – BTL DATA

RISE TIME TO CREST IN MICROSECONDS

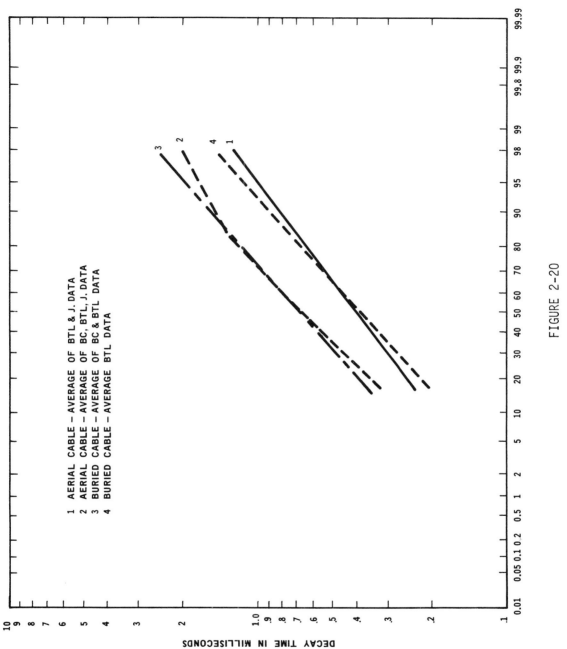

FIGURE 2-20

Distribution of Decay Times to Half Crest Values

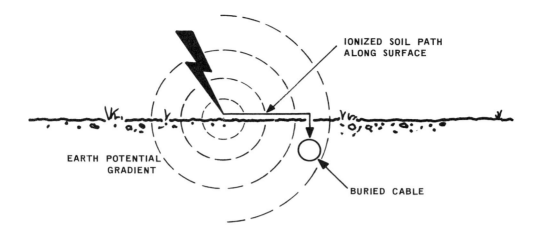

SOIL RESISTIVITY IN OHM— METRES	ARCING DISTANCE IN FEET
100	14
500	20
1000	26

FIGURE 2-21

Arcing Distances to a Buried Cable
Produced by an Average Stroke (30 kA)

MAGNITUDE DISTRIBUTION OF STROKE CURRENTS.   A magnitude distribution
of stroke currents to buried metallic objects is given in Figure 2-6.
The waveshapes of currents measured on the shield of a buried cable
are given in Reference 2.   The study of shield currents was conducted
for three lightning seasons in an area having nominal annual
incidence of 50 TD.   To obtain such data, it was necessary to make
simultaneous readings at about 20 points along a 21.5-mile section
of cable.   During these tests, 108 TD were observed in the general
area of the test cable.   The approximate waveshapes of shield
currents derived from data obtained are given in Table 2-13.

TABLE 2-13

Approximate Shield Current Waveshapes

STROKE NO.	EXTRAPOLATED CREST CURRENT (KILOAMPERES)	EXTRAPOLATED MAX. CHARGE (COULOMBS)	DERIVED TIME TO HALF VALUE (MICROSECONDS)
1	30	7	430
2	20	4	170
3	15	14	950
4	16	3.6	190
5	16	4	240
6	50	11.2	190
7	20	12	580
8	14	4	240
9	14	3.8	180
10	10	6.4	620
11	20	8	370
12	35	15	530
13	70*	11.2	130
14	50	8.8	140
15	12	12	1000

*   Measurement obtained at stroke point.

The data sample is small, but there nevertheless appears to
be a rough inverse correlation between current magnitude and decay
times.   The three exceptionally large stroke currents have decay times

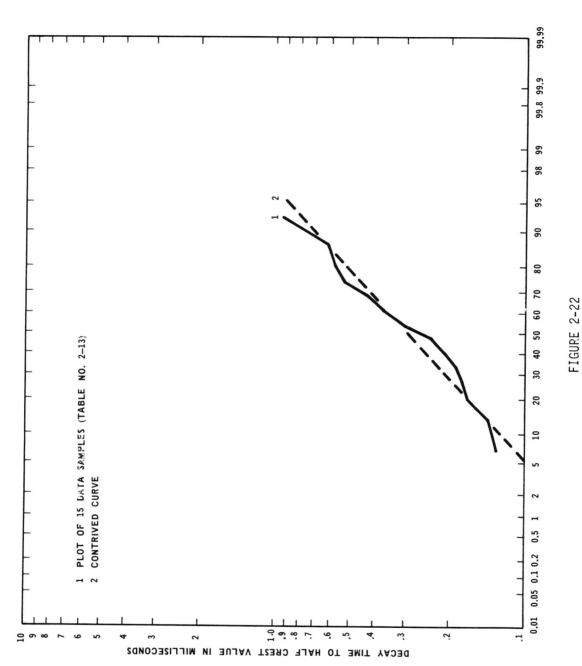

FIGURE 2-22

Decay Time of Surge Current in the Shield of

to half value of 130 to 190 μs, and 75 percent of the smaller stroke currents have longer decay times. A very rough plot of decay time distribution is given in Figure 2-22. Some liberty was taken in straightening out the curve on the assumption that the time distribution of the shield current wave will have approximately the same shape as that of the longitudinal voltage wave in core conductors.

SURGE VOLTAGE MAGNITUDES. Very little measured data is available on longitudinal surge voltages on core conductors (the reason for this is discussed under Aerial Exchange — Surge Parameters), and the little information available[11],[12],[20] does not distinguish between aerial and buried plant. This reflects the general assumption that the magnitude distribution in paired type cable does not differ greatly between aerial and buried construction. Buried trunk cable lends itself better to analytical predictions than does exchange cable, and Reference 2 offers the following expression for estimating core-shield voltage in such plant:

$$\text{Core Shield Voltage} = V_{c-s} \simeq KIR(\rho)^{\frac{1}{2}} , \qquad (2\text{-}7)$$

where

K is a waveshape factor, values of which are given in Table 2-14,

I = crest value of *total* stroke current in kiloamperes,

R = dc resistance per mile of shield,

ρ = soil resistivity in ohms-metre.

TABLE 2-14
Waveshape Factors

WAVESHAPE OF SHIELD CURRENT	WAVESHAPE FACTOR (K)
10 × 130	3.2
15 × 195	3.9
20 × 260	4.5
25 × 325	5.0
30 × 390	5.5

The wave front of current in the shield within the limits ordinarily encountered has negligible effect on the crest value of the core-shield voltage.

SURGE CURRENT MAGNITUDES.   Lacking specific information, the assumption has been made that the magnitude distributions of longitudinal
surge currents in core conductors are roughly the same as in aerial
cable discussed earlier in this subsection.

WAVESHAPE OF SURGES IN CORE CONDUCTORS.   Statistically (Figure 2-19),
the longitudinal voltage surges in buried cable appear to reach their
crest in somewhat less time than in aerial cable. From an engineering standpoint, the difference is not significant, except perhaps in
connected apparatus having high pass characteristics. The high frequency components associated with the arc discharge of protector gaps
may also require consideration in such cases.

     The TAC survey shows no significant difference in decay times
between aerial and buried cables. However, the decay times reported
are substantially longer than those obtained in the BTL studies. The
reason for this difference in reported data is not apparent. From
an engineering standpoint, it is important to note that the 10 × 1000
test wave developed from the BTL data appears to be adequate for design testing. For coordination with 3-mil carbon block protectors,
a minimum crest magnitude of 800 peak volts has been used and has
produced good field results. More recent thinking is in the direction of shortening the decay time of this test wave. Information recently received from the BTL protection group indicates that they
have, for most cases, shortened the half value decay time to 600 microseconds and increased the amplitude to 1000 peak volts to reflect the
higher initial surge sparkover of Western Electric 3-mil carbon blocks
of present manufacture.

2.2.12 Summary of Surge Current Parameters

OPEN WIRE.   The maximum current appearing in a line conductor will
depend on how much voltage the dielectric of the line will support
and on the conductor surge impedance. Opinions vary on the breakdown strength of open wire lines. The South Africans[10a] give maximum values of 70 kV for subscriber lines and 120 kV for trunk lines,
but state that in practice these values are probably lower. Sunde[5]
estimates flashover at voltages as high as 500 kV. Probable currents
based on these different estimates of line breakdown voltage for
points sufficiently removed from a stroke point for the line to assume a typical surge impedance are given in Table 2-15.

     There are no measured data on the waveshape of such surges
but, since surge impedance is in the nature of a resistance, it seems
reasonable to assume that surge current will have about the same waveshape as the voltage.

TABLE 2-15
Estimated Currents for Different Flashover Voltages

LINE FLASHOVER VOLTAGE	Z_0 OF 10 WIRE LINE	CONDUCTOR SURGE CURRENT (AMPERES)
70 kV	200	350
120 kV	200	600
500 kV	200	2500

Surge current entering a subscriber station located very
close to a point where lightning strikes the line may be higher than
the values in Table 2-15. As a practical check, let us review the
surge fusing values of 7-ampere station fuses that are occasionally
operated by lightning.

TABLE 2-16
Surge Fusing Characteristic of 7A Station Fuse

ASSUMED DECAY TIME (MICROSECONDS TO HALF VALUE)	FUSING CURRENT IN AMPERES
100	4000
150	3000
200	2450
300	1900

CABLES. This topic is considered under two headings:

1) Terminal Locations and Intermediate Repeaters. Surge current magni-
    tudes were considered, and from the available data and analytical
considerations it seems unlikely that surge currents will exceed
about 100 peak amperes. Waveshape distributions are probably similar to the observed longitudinal surge voltage waves.

2) Subscriber Stations. Surge currents discharged through protectors
    at subscriber stations may be higher than those experienced in trunk
cables and at terminal locations such as central offices. Conditions
that may produce higher discharge currents are:

    a) Stroke current on a cable shield flowing via conductor
        breakdown in terminals or face plate flashover to drop
    wires and then to ground through the station protectors.

    b) A stroke, either directly to or in the vicinity of a
        station ground, that produces high earth potential.
    Current will then flow through the station protectors to
    the cable.

Surge current measurements have been made on station drops served from aerial cable in a high exposure area. The highest amplitude in a relatively small amount of data was 2200 peak amperes. This was the total current in the drop wire so, assuming a division of current between the two conductors, the crest current per protector gap would be 1100 amperes. In Reference 5, it is stated that surge currents to cables over drops produced by high magnitude strokes near station grounds are likely to be in the range of 1000 to 2000 peak amperes. High magnitude currents observed close to the stroke point will be of much shorter duration than those at points where the surge has propagated through a substantial length of cable.

2.2.13 Parameters for Simulation Testing

Generalized surge parameters are commonly employed with good success for design testing of apparatus. Where it is known that actual operating conditions will significantly deviate from generally assumed environmental conditions or economic considerations are a forcing factor, apparatus designers may wish to use test values unique to the particular application. This text provides an analytical background and data base for such decisions.

Suggested generalized test values based on consideration of measured data field experience are outlined in Table 2-17.

2.2.14 Coaxial Cable (20 Tubes)

GENERAL. The data presented was obtained at two main stations on an L4 carrier route. The cable contained 20 coaxial tubes, was buried in soil ranging from about 300 to 1300 ohm-metres, and had two shield wires located 2 feet above it. Each test section was 8 miles long (4 repeater sections). The magnitude and waveshape of the induced lightning surges were measured on the inner conductor of a spare tube.

SURGE VOLTAGE MAGNITUDES. A distribution of measured peak voltages is shown in Figure 2-23. Crest voltages observed on the coaxial conductors were substantially less than on paired cable conductors. At the 95 percent distribution point, the respective values are 220 versus 400 volts. This would compare with approximately 1000 volts in paired cable if protectors were not used.

WAVESHAPES OF SURGES ON INNER COAXIAL CONDUCTORS. Distributions of rise and decay times are given in Figures 2-24 and 2-25. The data appear to be reasonably reliable from about the 2 percent to the 95 percent distribution points. Time to crest at the 2 percent

TABLE 2-17
Lightning Surge Test Parameters

APPLICATION	RISE TIME IN μs	DECAY TIME TO HALF CREST VOLTAGE[1]		PROTECTOR DISCHARGE CURRENT PER GAP (PK. AMPERES)
		IN μs	(PK. VOLTS)	
OPEN WIRE PLANT				
1) Cable-open wire junctions	4	200	800 – 1000	1500
2) Subscriber Stations	2	100	800 – 1000	2500
CABLE PLANT				
1) Trunk	10	1000	800 – 1000	100
2) Exchange				
a) Central Office	10	1000	800 – 1000	100
b) Subscriber Stations	4	200	800 – 1000	1500

Notes:

1. Voltage values based on use of 3-mil carbon block protectors. If lower sparkover devices are to be employed, a commensurate reduction in test voltage may be made.

2. There is only a small probability of exceeding the discharge current values given. Higher discharge currents will probably cause permanent protector grounding, but protection will be maintained.

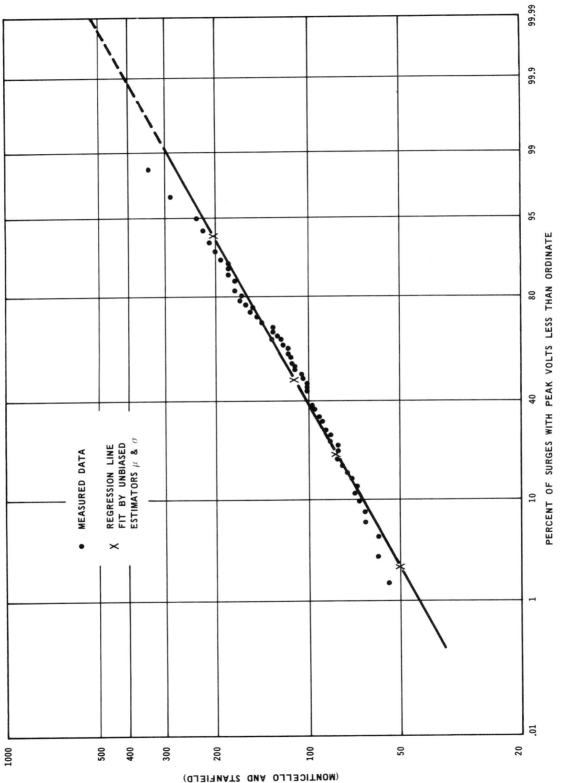

FIGURE 2-23 Probability Distribution of Measured Peak Voltage of Coax 20 Cable (from BTL)

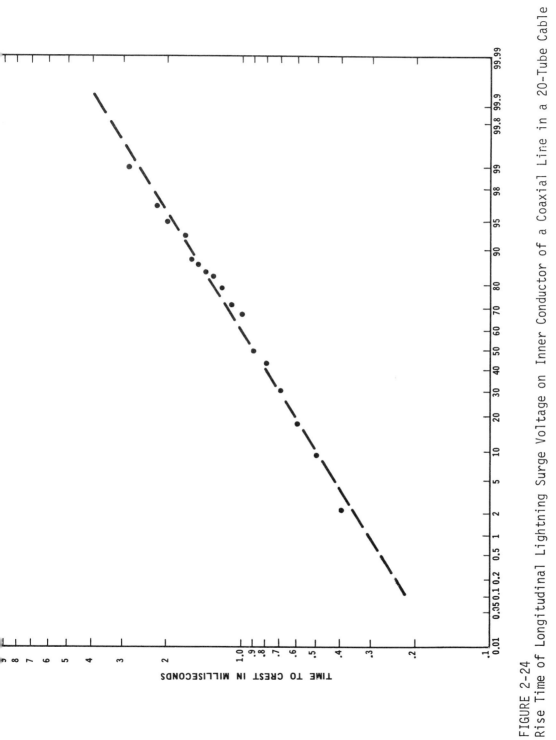

FIGURE 2-24
Rise Time of Longitudinal Lightning Surge Voltage on Inner Conductor of a Coaxial Line in a 20-Tube Cable

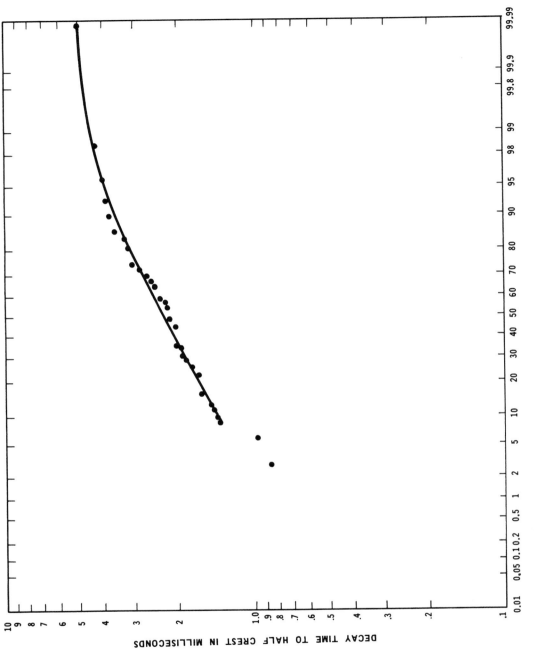

FIGURE 2-25 Decay Time to Half Crest Value of Longitudinal Lightning Surge Voltage
on Inner Conductor of a Coaxial Line in a 20-Tube Coaxial Cable

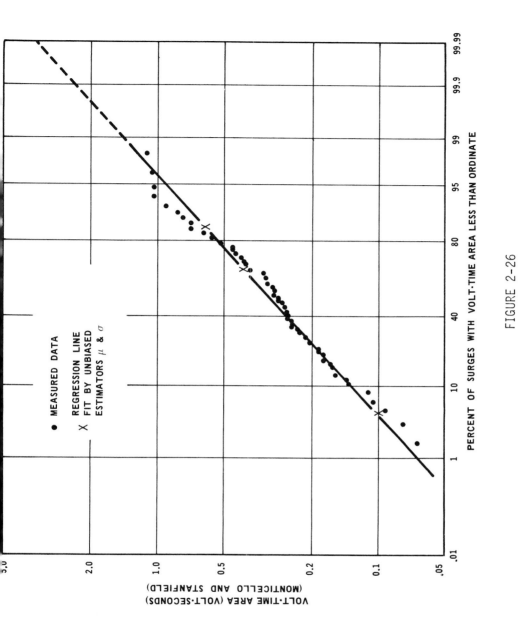

FIGURE 2-26

Probability Distribution of Measured Volt–Time Area
of Coax 20 Cable (from BTL)

point is roughly 360 microseconds, and decay time to half value at the 95 percent point is approximately 4000 microseconds. Comparable values for paired cable conductors are about 20×1000.

Data from all the cable studies reviewed support the general observation that longitudinal voltage magnitudes decrease, and rise and decay times increase, when increased shielding is provided in the cable structure. This behavior should be reflected in the Bell Canada LD-4 cable, which employs a double shield over the core.

VOLT-TIME RELATIONSHIPS. Volt-time values of the surges recorded in the BTL Coaxial Study are given in Figure 2-26. At the 95 percent distribution point, the volt·second value is 0.9. A comparable value for paired cable is approximately 0.56 based on an amplitude of 400 volts and decay time to half value of 1000 µs. The product of voltage squared and time in seconds (v·v·t) is roughly 137 for the coaxial and 114 for paired cable.

2.2.15 Crushing of Buried Cables by Lightning

INTRODUCTION. Buried air dielectric coaxial cables are especially susceptible to physical crushing by pressures associated with the vaporization of moisture in the soil due to direct lightning discharges to the shield. Photographs of a typical sheath dent in a coaxial cable and its effect on the coaxial tubes are given in Figures 2-27 and 2-28 respectively. Damage may range from a degree of deformation that produces corona or permits power voltage arc-over to complete tube collapse (metallic short circuit).

Crushing of buried paired cable occurs, but the compression rarely damages the conductor insulation. Laceration of paper insulation has been observed where the sheath has been severely crushed, but PIC is practically immune to degradation of conductor insulation from such pressure. It is only in buried air dielectric coaxial plant that crushing is a problem requiring consideration.

In the larger size multitube coaxial cables, typically used as broad band carriers, failure of insulation from electric stress is unlikely to occur unless the normal dielectric strength has been degraded due to some physical damage or poor workmanship.

FIELD TROUBLE EXPERIENCE. Experience to date has been with AT&TCo L-carrier cables that employ lead sheath construction. A BTL memorandum[21] (December 29, 1965) states that lightning crushing occurs at a rate of 2.6 cases per 1000 miles per year. In considering the practical implications of this figure, the following factors should be recognized:

1) Some amount of the total cable mileage has shield wires in a variety of configurations that provide a degree of

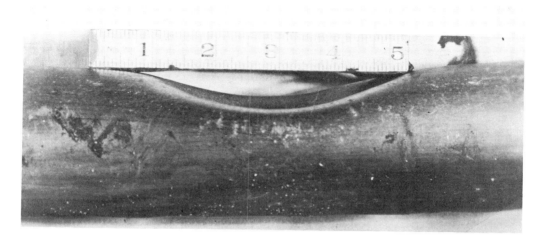

FIGURE 2-27

Coaxial Cable Crushed by Simulated Lightning

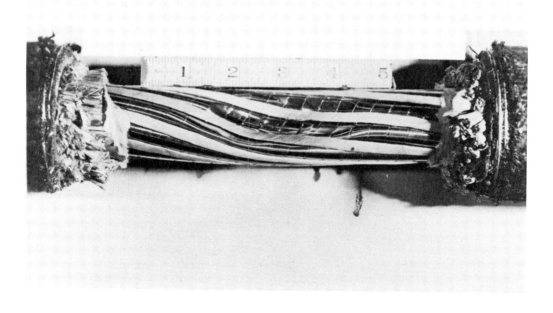

FIGURE 2-28

The Core Related to the Sheath Shown in Figure 2-27

protection. The mileage so protected is not specified. It is known, however, that shield wires were initially placed with very little of the early coaxial plant. Some shield wire protection was subsequently installed in specific sections, but it probably did not exceed about 700 miles. With the introduction of 'hardened' plant, the use of two-shield wires became standard practice. At the time that the above-mentioned crushing rate was derived, a rough guess would be that about 50 percent of the buried coaxial plant had some form of shield wire protection.

2) The crushing environment is quite nonuniform. In addition to lightning incidence, soil conditions, especially moisture, are very critical.

3) The size and type of cable construction.

4) Power voltages in the newer systems are higher than originally used, which increases susceptibility to service interruption for a given degree of cable deformation.

NATURE OF STROKE CURRENT PRODUCING CRUSHING. Field inspection of crushed cables shows that usually there is only minor evidence of fusing of the lead shield. This observation and subsequent laboratory investigation have led to the use of the terms 'cold' and 'hot' lightning. These terms were not original, however, since they had been used previously with respect to explosive-type damage to structures. It has been observed frequently that trees and wood structures can be extensively damaged in an explosive manner, yet there is little or no evidence of combustion. These effects were concluded to be the result of the delivery of a large amount of electricity (coulombs) in a relatively short stroke time and its interaction with moisture in the wood fibers.

In the case of cable crushing, evaluation of variables has been accomplished with good success through simulated lightning tests in which cable specimens were buried in soil contained in large explosion proof boxes. In the initial study of the major variables associated with cable crushing[22], specimens of plain lead sheath paired cable were used because of the simple structure and homogeneous character of such cable. Also, it was somewhat easier to crush, which permitted greater range of test current amplitudes and durations. Subsequent tests with coaxial cable specimens tended to confirm that the relationships between the fundamental variables associated with crushing obtained in the paired cable tests are generally applicable. The more important factors established by these tests are outlined below:

1) For a given condition, crushing increases with the coulombs delivered to the shield via the arc.

2) The degree of denting per coulomb is greater on shorter duration surges. It appears that the heat produced by the arc is more effectively converted to denting pressure when applied within a relatively short time. It was found by test that the denting per coulomb ratio with a current impulse having a decay time to half crest value of 20 microseconds decreased at a much slower rate.

3) In these tests[22], crushing was produced with impulses having a decay time to half crest in the range of about 13 to 150 microseconds. With decay times up to about 40 microseconds, little evidence of fusing was observed on the lead shield. When the decay time was extended to 153 microseconds, fusing was considerably more evident.

PHYSICAL FACTORS. Physical factors having a significant effect on the degree of crushing are discussed below in their approximate order of importance.

1) Amount of water in the soil: Crushing is not significant in dry soil but increases with the moisture content until a maximum is reached in saturated soil. This relationship, established by test, has been confirmed by field observations. Crushing occurrence has been most severe in swampy areas or where the water table has reached the depth of the cable and produced water saturated soil, as in the case, for example, in parts of Florida and Mississippi. Conversely, little crushing has been experienced in the southwestern part of the U.S.A., where the soil is typically dry.

2) Cable construction: It has been observed[22] that plain lead sheath cable of constant thickness will dent more easily as the diameter is increased. Also, as would be expected, increasing the lead thickness with a given core diameter makes a cable more resistant to crushing. In practice, however, the thickness of lead sheath usually increases with diameter, so these two factors tend to compensate. Tests indicated that the extruded polyethylene core insulation in Lepeth cable makes a modest contribution to the structural strength of the sheath. Tape armor, within the range customarily used, provides only small additional strength. However, light wire armor makes a substantial contribution against crushing. It has also been observed in tests that compression of the core structure provides significant resistance to further crushing as deformation proceeds.

3) Number of tubes: As will be noted in the next subsection, for the types of cable construction employed by AT&T, the physical size of the sheath dent increases with the number of tubes, but the ratio of damaged tubes to total tube count decreases.

4) Soil Composition: In tests, the composition of the soil
was not found to be a major factor. Compaction, as would
be expected, in practice increases the degree of crushing.

TUBE CRUSHING VERSUS CURRENT MAGNITUDES. Data from Reference 21,
showing relative tube damage in types of cable of present interest,
are plotted in Figure 2-29. Tube damage was based on a 2000 V (rms)
'Hypot' test. A cable code description is given in Table 2-18. The
8-tube cable in this figure is significant from a control standpoint
in that the crushing trouble rate previously quoted under Field
Trouble Experience is based to a considerable extent on this type of
cable. Likewise, data for the additional cable types provide a
basis for relating field experience and evaluating, by laboratory
test, procedures and cable structures anticipated for future use.
Table 2-19 gives a comparison between the number of tubes that
appeared to be damaged on the 2000 V Hypot test and tubes indicating
damage on a 1000 V Hypot test under comparable conditions.

As a matter of related interest, tests were also made on
specimens of hollow steel pipe. These test specimens were seamless,
4½" inside diameter of 13 ga. (0.095") SAE1015 steel. The crushing
observed gives further insight into the severe pressures associated
with this phenomena (Figure 2-30). This figure also shows the
degree of external denting experienced by coaxial cable under comparable test conditions.

TABLE 2-18

Cable Code Description

1.	PCM	- CF10121 (Experimental PCM Cable with Tolpeth K sheath, O.D.2.28 in., 5 pairs and 18 conductors of 19 gauge, 18 coaxials of .375 in. dia.)
2.	20 COAX	- CA3091 made according to N.E. specification CF10114 (Lepeth PJ sheath, O.D.3.05 in., 4 pairs of 16 gauge plus 1 quad, 41 pairs and 10 conductors of 19 gauge, 20 coaxials of .375 in. dia.)
3.	12 COAX	- CA1958F (Lepeth PJ sheath, O.D.2.69 in., 4 pairs of 16 gauge plus 7 quads and 12 conductors 19 gauge, 12 coaxials of .375 in. dia.)
4.	8 COAX	- CA1264 (Lead TCP sheath, O.D.2.03 in., 8 pairs of 16 gauge plus 2 quads and 18 conductors of 19 gauge, 8 coaxials of .375 in. dia.)

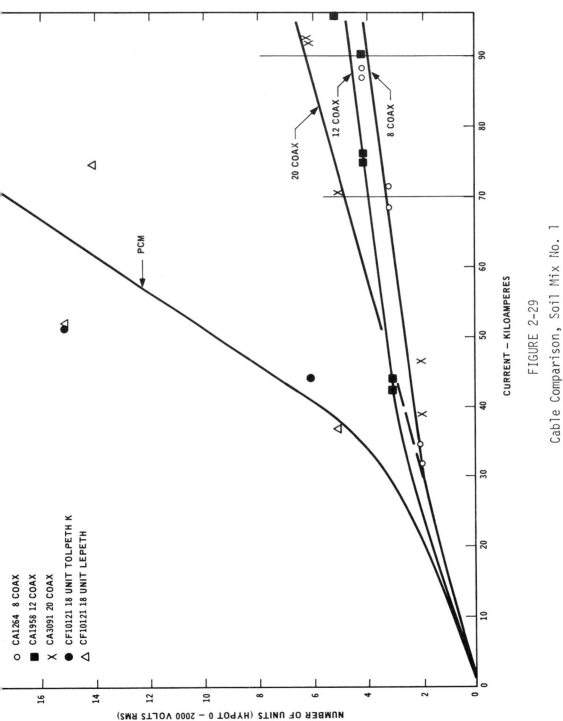

FIGURE 2-29
Cable Comparison, Soil Mix No. 1

TABLE 2-19

A Summary of Dielectric Strengths

TYPE OF CABLE	RELATIVE CURRENT MAGNITUDES	NO. OF TUBES DAMAGED	
		< 2000 V (rms) Hypot	< 1000 V (rms) Hypot
1264 (8 Tube)	Low	2	2
1264 (8 Tube)	Low	2	1
1958 (12 Tube)	Low	3	2
1958 (12 Tube)	Low	3	3
10114 (20 Tube)	Low	2	2
10114 (20 Tube)	Low	2	1
PCM (Tol)	Low	6	5
PCM (Lep)	Low	5	4
1264	Medium	3	1
1264	Medium	3	3
1958	Medium	4	3
1958	Medium	4	3
10114	Medium	5	4
PCM (Tol)	Medium	15	14
PCM (Lep)	Medium	15	12
1264	High	4	4
1264	High	4	3
1958	High	4	4
1958	High	5	3
10114	High	6	4
10114	High	6	4
PCM (Tol)	High	18	18
PCM (Lep)	High	14	14

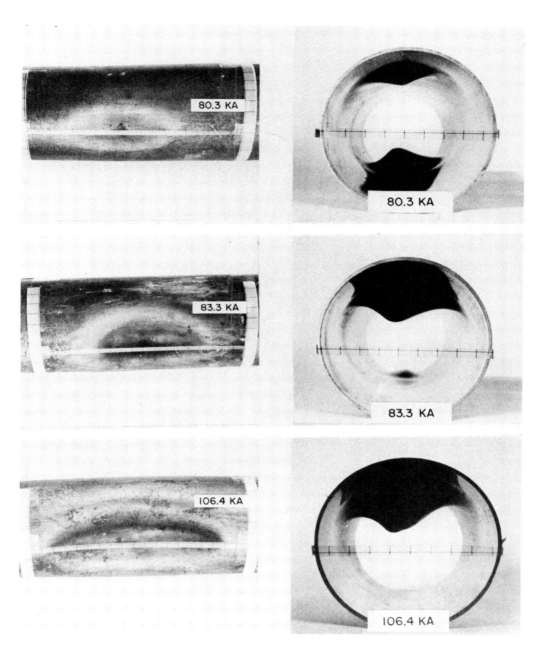

FIGURE 2-30

PITTSFIELD SURGE TESTS

40% Clay and 60% Sand Test Medium
Seamless Steel Pipe (4½ in. I.D.)

2.2.16 Consideration of Multiple Strokes

A lightning stroke may consist of one or more components of surge currents that discharge through the same path in the atmosphere. This is not a well resolved subject, but some available information indicates that there are substantially more components in strokes to tall objects than, for example, to a communication line. It has also been observed that only about one third of the strokes to power transmission lines contain multiple components, and, on the average, these have only one or two components in addition to the initial discharge. The time interval between peaks ranges from about 0.05 to 0.1 second.

Automatic cathode ray oscillographs used to provide the type of data presented in this text do not cycle with sufficient speed to record multiple strokes, and the time base was set to resolve only the initial surge. A very long sweep time would be required to investigate multiple components, in which case only the peaks would be recorded, and resolution would be too poor to characterize waveshapes.

Laboratory testing is conducted with a single impulse, probably because of the difficulty of developing a series of high energy impulses closely spaced in time. There has been no pressure, however, to change this procedure because long years of experience indicate that both power and communication apparatus tested with a single impulse will coordinate satisfactorily with the actual environment. There should be little doubt that a single impulse is adequate for testing of insulation, especially when the test wave approaches a most severe case.

Occasionally there is some speculation concerning the adequacy of this procedure in the case of apparatus containing semiconductor components. Field experience has established the practical effectiveness of testing such apparatus on a single impulse basis. The probable reason for this good experience is related to the basic protection provided for exposed apparatus. It is assumed that apparatus design tests have been conducted on a full wave basis with voltage amplitudes of 800 to 1000 peak volts. In practice, most protectors will sparkover at 500 to 600 peak volts or less (especially gas tubes). When the protectors discharge, input voltage to the apparatus is reduced to the range of 25 to 50 peak volts. This usually occurs rapidly so, for most of the surge, relatively little energy reaches the apparatus. Full wave exposure only occurs when the surge voltage is less than required to operate the two protector gaps. Since design tests are conducted on a full wave basis at considerably higher voltage amplitudes, there is a substantial protective margin available to compensate for possible effects of subsequent stroke components.

2.3 REFERENCES

1. Special Issue on Lightning Research, Journal of the *Franklin Institute*, June 1967.

2. H.M. Trueblood, E.D. Sunde, "Lightning Current Observations in Buried Cable", *BSTJ*, Vol. 28, April 1949, pp. 278-302.

3. E.J. Workman, "The Production of Thunderstorm Electricity", Journal of the *Franklin Institute*, Vol. 283, June 1967, pp. 540-551.

4. K. Berger, "Results of Research on Mount San Salvatore", Journal of the *Franklin Institute*, Novel Observations on Lightning Discharges, June 1967, p. 502.

5. E.D. Sunde, "Earth Conduction Effects in Transmission Systems", *Dover Publications*, New York.

6. C.F. Wagner, G.D. McCann, "Lightning Phenomena", Electrical Transmission and Distribution Reference Book, *Westinghouse Electric* Corp. Chapter 16, pp. 542-577.

7. E.D. Sunde, "Earth Conduction Effects in Transmission Systems", *Dover Publications*, New York, pp. 268-298.

8. J.H. Hagenguth, J.G. Anderson, "Lightning to the Empire State Building", *AIEE* Transactions, Vol. LXXI, August 1952, p. 52.

9. C.F. Wagner *et al*, "Field Investigations on Lightning", *AIEE Transactions*, Vol. LX, 1941, p. 1222.

10. National Bureau of Standards Handbook #46, 1952, U.S. Department of Commerce.

10a. D.P.J. Retief, I.C. Ramsay, C.F. Boyce, "The Protection of Open Wire Communication Systems from Lightning Damage and Interference with Particular Reference to South Africa", *Monograph*, Associated Scientific and Technical Societies of South Africa, June 1955.

11. D.W. Bodle, P.A. Gresh, "Lightning Surges in Paired Telephone Cable Facilities", *BSTJ*, Vol. XL, March 1961, pp. 547-576.

12. E. Bennison, A.J. Ghazi, P. Ferland, "Lightning Surges in Open Wire, Coaxial and Paired Cable", *IEEE*, International Conference in Communications, June 1972. (This paper presented information from the TCA Lightning Survey, TAC-3.)

13. A.K. Waldorf, "Experience with Preventive Lightning Protection on Transmission Lines", *AIEE* Transactions, Vol. LX, 1941, p. 249.

14. E. Hansson and S.K. Waldorf, "An Eight Year Investigation of Lightning Currents and Preventive Lightning Protection on a Transmission System", *AIEE* Transactions, Vol. LXIII, 1944, pp. 251-258.

15. D.W. Bodle, "Relationship of Grounding and Bonding to the Effectiveness of Lightning Protection Devices", *IEEE* Conference Record, 1970, Annual Meeting.

16. R.H. Card, "Earth Resistivity and Geological Structure", *Electrical Engineering*, LIV, November 1935, p. 1153.

17. J.R. Eaton, "Moisture, Temperature and Soil Resistivity", *Electrical World*, CLX, August 1941, p. 602.

18. J.R. Eaton, "Impulse Characteristics of Electrical Connections to Earth", *General Electric* Review, XLVII, October 1944, pp. 41-50.

19. "Lightning Arresters for Alternating Current Power Circuits", *ANSI*, C62.1.

20. Yukinori Ishida, "Induced Lightning Surges in Paired Telephone Cables", Review of the *Communication Laboratories*, Vol. 20, No. 3-4, March-April 1972.

21. R.D. Gunther, "Simulated Lightning Crushing Tests of Multicoaxial Cable", *BTL* Report of December 1965.

22. D.W. Bodle, "Final Report on the Results of the Pittsfield Cable Crushing Tests", *BTL* Report of February 1954.

3. POWER INTERFERENCE
(Under Abnormal Conditions)

3.1 INTRODUCTION

The Power network is an important environmental factor affecting communication facilities. Steady state induction at fundamental and harmonic frequencies often causes signal interference (noise) in communication systems.

Paralleling telephone wires and cables may also be exposed, during power line phase-to-ground faults, to induced voltages of sufficient magnitudes to damage connected apparatus and to be a shock hazard to field personnel.

In joint-use aerial construction and at Power-Telephone cross-overs, there exists the possibility of direct contacts.

Although good plant practices minimize such occurrences, direct contacts must still be recognized as a significant factor because of the associated high levels of voltages and currents produced by them in the telephone plant.

Practices employed by the power industry are subject to continual change which contributes to the difficult task of characterizing power network from a protection standpoint.

Power systems can adversely affect telecommunications plant in a number of ways. For example, through resistive or inductive coupling, they can produce voltages that are hazardous to personnel and the public. They can also cause damage to plant and property as a result of overstressed dielectrics or overheated elements. Furthermore, the frequency spectrum produced by power systems interference is such that it can endanger the quality and reliability of the telecommunications transmissions.

A power system will interfere with telecommunications plant if the system generates a net electromagnetic field that couples to the telecommunications plant with sufficient intensity and with such a frequency content that either the physical integrity of the circuit is endangered or the transmission of information is impaired.

The purpose of this section is to characterize the elements of power systems that produce interference in communication circuits under abnormal conditions. Attention is directed to those aspects of power generation, transmission, and distribution that produce interference hazardous to personnel and plant. Steady state interference and disturbances in the noise domain will not be dealt with here since a separate study is being carried out in these areas.

3.2 SCOPE

Couplings between power and communications systems are rarely present in one form alone. Usually there will be a combination of the three forms of coupling, i.e., resistive, capacitive, and inductive. However, one form of coupling is usually predominant, and hence the combined effects of the different couplings can be analysed with relative ease. In this section, separate analyses and statistics are presented for each of the following forms of coupling:

a) Capacitive coupling: Because most of the existing and proposed telephone plant is screened to earth, capacitive coupling is only described analytically.

b) Resistive coupling: The most recent statistics for power contacts in the Western, Eastern, and Central areas of Bell Canada are presented. Data on power station and tower ground potential rises are also given.

c) Inductive coupling: The data here have been gathered from existing route layouts of the telephone and power systems. These data cover most of the toll network interface with power transmission systems, and two rural exchange trunk areas that interface with the subtransmissions and distribution networks in Quebec and Ontario.

The following electrical parameters are covered: fault current distributions, fault durations, frequency of occurrence, and X/R ratio* distributions per voltage class. The geographical parameters discussed include earth resistivity, separation, and exposure lengths.

3.3 CHARACTERISTICS OF POWER SYSTEMS AND THEIR EFFECTS ON TELECOMMUNICATIONS PLANT

3.3.1 General

Electric power is produced by generating plants at a relatively low voltage, which must be stepped up to reduce the losses incurred in transmitting it to large load centers (see Figure 3-1). At the load centers, power is apportioned in bulk through the subtransmission network, transformed to primary feeder voltages, and channelled to distribution transformers. The final stage in the system is the distribution to customers of 110/220-volt power via distribution transformer secondaries. (It should be noted that, because of the evolution of power networks, circuits that were originally designed for one function are now performing another; for example, circuits that were built for transmission some years ago and are still being used would, by today's standards, be classified as subtransmission.)

* The ratio of the reactive and resistive components of an electrical network.

Interference can take place when a net electromagnetic field is present due to a 'permanent' or a 'temporary' disturbance on a power system. Permanent interference is caused by electrical or geometrical unbalances of the power system. Temporary interference is caused by faults or switching transients.

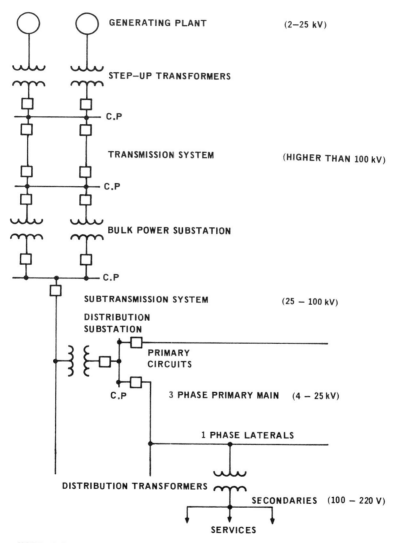

FIGURE 3-1
The Functional Components of an Electric Power System
(After Reference 1)

The most frequent and severe faults from the point of view of a communications engineer are line-to-ground (L/G) faults*. It is therefore important to know the response of power systems to such faults when the other parameters are changed. The following subsections present a description of power system parameters that can influence the degree of unbalance, and also a brief description of the significance of these parameters in terms of protection.

3.3.2 Circuit Connection

DELTA (MESH) CIRCUITS. In a delta-connected circuit, line-to-ground faults generate very low residual currents (ground return currents), since the only return path to the source is via the distributed capacitance-to-ground of the sound (unfaulted) phases.

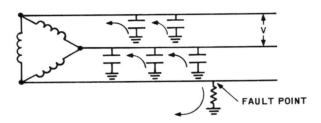

FIGURE 3-2

Line-to-Ground Fault on a Delta System

The approximate magnitude of the fault current is given by the emperical relation

$$I = 5 \times 10^{-3} \ V \cdot L \ [A_{rms}]$$

where V = phase-to-phase voltage (in kV), and
       L = line length (in miles).

Because of their low values, L/G faults do not disturb the operation of a delta circuit, but the delta circuit does have some disadvantages. With existing relaying techniques, rapid detection and clearing of L/G faults is difficult to achieve, and this can result in safety hazards and overstress of the insulation of the sound phases. Adequate joint-use coordination can only be achieved with sacrifice in economy. Earth current detection must be employed on the delta power system and the telephone plant must be effectively grounded. Indirect sharing is possible, however, when there is a grounded

*Single-phase-to-ground faults.

circuit between the communication facility and the delta line (Umbrella Effect). Because of the stress between the sound phases and ground, line-to-ground faults on delta systems can degenerate to two-phases-to-ground faults, causing a substantial residual current to flow, and possibly affecting telecommunications lines due to magnetic coupling.

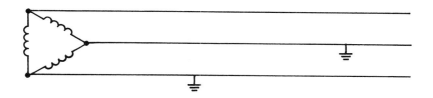

FIGURE 3-3

Two-Phases-to-Ground Fault on a Delta System

UNIGROUNDED Y CIRCUITS AND DELTA CIRCUITS WITH GROUNDING BANKS. These configurations are equivalent as far as the problem of interference is concerned. L/G fault currents in such systems are lower than those in equivalent multigrounded systems, and thus positive disconnection of a faulted line does not occur as quickly. Joint use with uni-grounded circuits is avoided for the same reasons mentioned earlier for delta circuits.

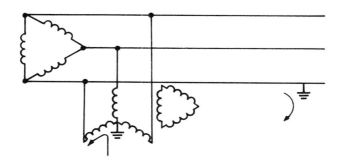

FIGURE 3-4

Star-Delta Grounding Transformer on a Delta Circuit

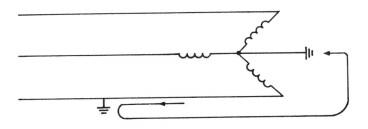

FIGURE 3-5

Unigrounded Y Circuit

MULTIGROUNDED Y CIRCUITS. Multigrounded Y circuits have the highest
residual currents for L/G faults because of the reduced zero-sequence
impedance. However, prompt de-energization of such lines and improved
grounding of the telephone plant by interconnecting it with the
multigrounded power lines facilitate good protection coordination.

Since positive disconnection of multigrounded Y circuits from
L/G faults is certain, they are considered safe for joint use up to
44 kV in Bell Canada. Although known coordinating techniques may be
employed to facilitate safe joint use at these higher voltages,
economic aspects are important in determining the feasibility of
such practices.

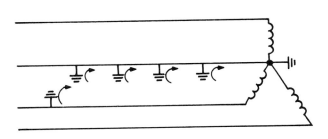

FIGURE 3-6

Multigrounded Y Circuit

3.3.3 Load: Electrical and Geometric Unbalances

A power line may be perfectly balanced electrically and geometrically, but differences in spacing between each phase of the power line and the communication line result in different degrees of coupling. This creates a net emf coupled either inductively or resistively. In existing subtransmission and distribution circuits, geometric unbalances are unlikely to present a threat where the power and communications lines are separated by more than 200 feet.

The balance of loads is an important factor in the operation of a system. In transmission circuits, the balancing of loads is relatively easy to achieve, whereas geometric balance, even after transposition of phase wires, is much more difficult.

In distribution circuits, the reverse is true. Although auxiliaries, services, and branch feeders are arranged so that loads are balanced at the source (e.g., at a distribution transformer), the presence of tappings and branchings (which are very common in distribution circuits) means that the load at a point remote from the source can be completely out of balance. Figure 3-7 illustrates the effect of branching; note that the residual permanent current in Zone 2 is the full phase load.

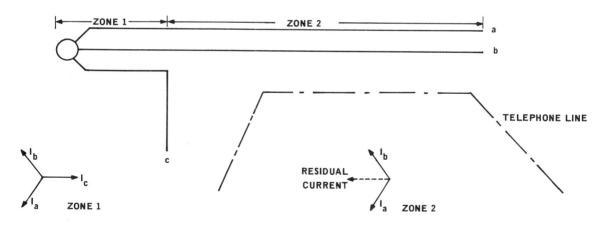

FIGURE 3-7

The Effect of Branching on a Distribution Circuit

HARMONICS. Interference to communication circuits can be caused by harmonics of the basic 60 Hz component of a power wave. These

TABLE 3-1
A Summary of Electrical Features of System Grounding
(After Reference 3)

	GROUNDING SCHEME			
	UNGROUNDED	EFFECTIVELY GROUNDED	REACTANCE GROUNDED	RESISTANCE GROUNDED
APPARATUS INSULATION	Fully Insulated	Lowest	Partially Graded	Partially Graded
FAULT TO GROUND CURRENT	Usually Low	Maximum value rarely higher than three-phase short circuit current	Cannot satisfactorily be reduced below one-half or one-third of values for solid grounding.	Low
SAFETY FROM VOLTAGE GRADIENT CONSIDERATIONS	Usually good, but not fully dependable because of possibility of simultaneous fault on another phase	Gives greatest gradients, but not usually a problem where continuous ground wires are used.	Slightly better than effective grounding	Better than effective or reactance grounded.
STABILITY	Usually unimportant	Lower than with other methods but can be made satisfactory by use of high speed relays and circuit breakers	Improved over solid grounding particularly if used at receiving end of system	Improved over effective grounding particularly if used at sending end of system
RELAYING	Difficult	Satisfactory	Satisfactory	Satisfactory
ARCING GROUNDS	Likely	Unlikely	Possible if reactance is excessive	Unlikely
LOCALIZING FAULTS	Effect of fault transmitted as excess voltage on sound phases to all parts of conductively connected network	Effect of faults localized to system or part of system where they occur	Effect of faults localized to system or part of system where they occur unless reactance is quite high	Effect of faults transmitted as excess voltage on sound phases to all parts of conductively connected network
DOUBLE FAULTS	Likely	Unlikely	Unlikely unless reactance is quite high and insulation weak	Unlikely unless resistance is quite high and insulation weak
LIGHTNING PROTECTION	(Ungrounded neutral service) arresters must be applied at sacrifice in cost and efficiency	Highest efficiency and lowest cost	If reactance is very high arresters for ungrounded neutral service must be applied at sacrifice in cost and efficiency	Arresters for ungrounded, neutral service usually must be applied at sacrifice in cost and efficiency
LINE AVAILABILITY	Will inherently clear themselves if total length of interconnected line is low and requires isolation from system in increasing percentages as length becomes greater	Must be isolated for each fault	Must be isolated for each fault	Must be isolated for each fault
ADAPTABILITY TO INTERCONNECTION	Cannot be interconnected unless interconnecting system is ungrounded or isolating transformers are used	Satisfactory indefinitely with reactance-grounded systems	Satisfactory indefinitely with solidly-grounded systems	Satisfactory with solidly- or reactance-grounded systems with proper attention to relaying

harmonics can be generated at a permanent load point by such effects as cyclic variations of reluctance in rotating machines, nonlinearities in the **magnetization** of ferromagnetic elements, and nonlinearities in rectifiers. Such interference represents a floor level that communications systems have to be able to survive and as such is classed in the noise domain.

3.3.4 Operating Voltages

Power line operating voltages directly control the severity of switching transients, corona, and electric induction. Since most cable plant has a grounded screen, the induced voltages are limited due to the shielding effect it provides, and the problem is mainly one of shock hazard rather than interference.

Corona is associated with ionization phenomena in the vicinity of power phase conductors and occurs when the electric field strength is high enough to cause breakdown of the surrounding air. Discharges give rise to current pulses, which cause power losses and radio interference. The latter phenomenon belongs in the noise domain. (For further information, see Reference 35).

At every switching operation on a power circuit, a variety of transient effects occur that can cause interference to communications systems. For example, a voltage reflection will occur on a line if the terminal impedance differs from the line surge impedance, and the voltage at the source can reach 200 percent of the normal operating value. Even with the symmetrical operation of a switch on a 3-phase line, a net voltage is produced that may be capacitively coupled to a communications line. (Induced voltages may also occur on the communications line through magnetic coupling, but these are usually of a small magnitude.)

CLASSIFICATION OF LINES BY VOLTAGE. Regardless of their actual operating voltage, power lines can be grouped into one of ten general classes when considering inductive interference (see Table 3-2). Each of these classes has a relatively narrow difference in interference levels due to resistive, capacitive, or magnetic coupling, for the different operating voltages it is associated with.

TABLE 3-2

Voltage Classes

RANGE (kV)	CLASS (kV)	TYPE
3.5 - 6.9 7 - 15	4 10	Distribution
16 - 30 31 - 50 51 - 89	25 45 65	Subtransmission
90 - 165 166 - 275 276 - 345 346 - 550 551 - 800	110 220 330 500 735	Transmission

3.3.5 Power Line Length

The length of a power line determines the exposure intensity and therefore affects the levels of both capacitively and magnetically coupled interference. For capacitively coupled interference, the line length determines the charge induced, and for magnetically coupled interference, it influences the induced field.

3.3.6 Line Construction And Grounding

The separation between conductors of a power system and the conductor-to-ground spacings are dictated primarily by the dielectric stresses at the operating voltage and the risk of contact between phases due to foreign objects or the effects of wind. Phase-to-phase separation only affects the permanent interference (noise).

The possibility of contacts between a transmission circuit and a foreign object can be reduced if the width of the right of way is increased. However, the maximum width that is economical is steadily decreasing because of the scarcity (and hence the increasing cost) of right of way.

The major cause of faults on transmission circuits is lightning but the rate of outage can be substantially reduced by the use of skywires, counterpoises, and tower footings, which decrease the rise of potentials and hence tne risk of flashover on phase insulators. Skywires and counterpoises share a portion of the return current, and thus reduce the net emf coupled magnetically.

3.4   Power System Protection[4,5,6]

The objectives of protection for the three main parts (generation, transmission, and distribution) of a power system are:

a)   isolation of the section that is in an abnormal condition,

b)   optimization of system reliability, and

c)   the efficient use of facilities.

An abnormal condition is one resulting from an overload, a system instability, or a fault. Such a condition can be persistent or nonpersistent, the former type being more characteristic of distribution circuits.

Protection measures for faults on a line are programmed for a succession of reclosing (re-energization) operations at a frequency dependent on the type of circuit in which the fault occurs. For example, reclosing sequences on transmission lines generally involve no more than two automatic attempts to re-energize the faulted line, since alternate routes are available and the fault may be persistent. On low-voltage circuits, where distribution is usually radial, up to four reclosing attempts may be made.

The duration of each energized interval (see Figure 3-8) includes the detection time, which can be intentionally delayed, and the operation time of the clearing device, which varies considerably. The de-energized intervals between successive reclosing operations differ with the circuit hierarchy and the reclosing policy of the distribution district. To effect selective clearing, power systems are divided into protective zones with different degrees of overlap (see Figure 3-9).

A typical protection system includes:

a)   detection and de-energization devices,

b)   communication channels between sensing devices on HV and EHV lines,

c)   arrangements of tripping sequences (security), and

d)   back-up protection.

Detection is usually based on one or more of the following:

a)   the difference between input and output powers,

b)   line impedance measurements,

c)   levels of phase or neutral currents,

d)   pressure,

e)   heat, and

f)   variation of frequency from 60 Hz*.

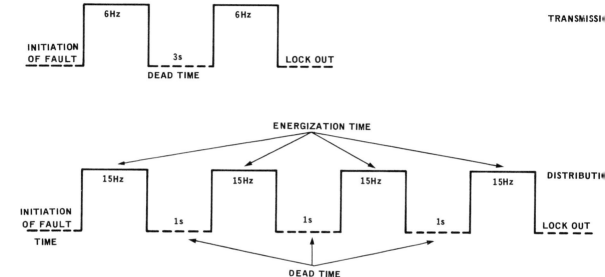

FIGURE 3-8

Examples of Reclosing Sequences for Typical
Distribution and Transmission Circuits

        Communication channels between sensing devices are only used
in conjunction with high-reliability lines (≥100 kV) and range in
complexity from a single pair of telephone wires to a radio channel,
depending on the system dependability requirements. On subtransmission and distribution lines, and on many 100 kV radial circuits,
communication channels are not used.

* All times shown in Hertz in this section are based on power system
  frequency of 60 cycles per second.

Electrically controlled coordination between fault detection
points (tripping and locking commands) is used on high-reliability
lines. In low-voltage circuits, coordination is achieved by timing.

All protection systems are provided with backup. These
backup devices rarely operate on high-reliability circuits. However,
their operation on subtransmission and distribution circuits is
more frequent.

Because of the reliability of the first stage of protection
and the negligible short delays due to tripping coordination and
communication, the overall duration of faults associated with transmission systems is primarily determined by the function times of
detection devices.

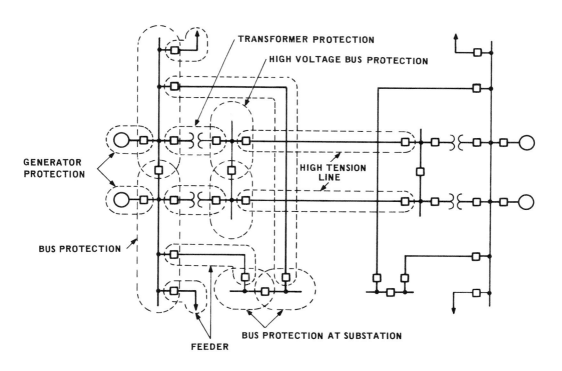

FIGURE 3-9

Typical Protection System, Showing Protective Zones

In distribution systems, the first stage of protection is usually a line fuse. Because of degradations due to external factors, the operation of these fuses is not always predictable. Thus, where there are persistent faults on a system, it is common practice to consider the effective duration of the faults as the operation time of the reclosers or relays.

Abnormal conditions that create a net electromagnetic field will cause magnetically coupled interference only if the flow path has an appreciable length. Localized faults only affect the ground potential rise. Protection systems can therefore be divided into two major categories, apparatus and bus protection (see below), and line protection (see 3.5).

APPARATUS AND BUS PROTECTION. In addition to fast-operating relays that protect against internal faults, supervision relays at apparatus and buses overlap the protection zone of the line. This second stage of protection has a long response time so that the fast relays of the line protection are given the first chance to clear the fault. However, where stability or fault damage would present a problem on the system, it is common practice to use high-speed relays for both primary and backup protection.

Two main types of detection are used to protect against internal faults:

    a) DIFFERENTIAL. This is the more common and operates on the difference between the input and output power at a transformer, generator, reactor, or bus. (See Figures 3-10, 3-11)

    b) THERMAL AND PRESSURE. Detection is based on the build-up of heat or pressure and these devices are used mainly at transformers and reactors. There are no intentional time delays with this type of protection.

99

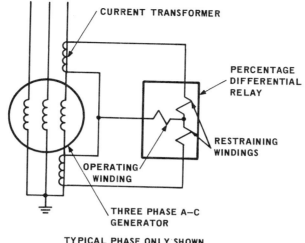

FIGURE 3-10

Connections for One Phase, Using a
Differential Relay for Generator Protection

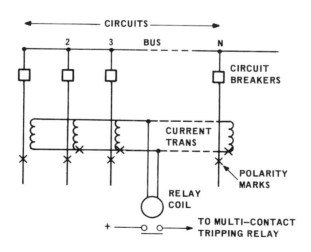

FIGURE 3-11

Bus-Differential-Current Relay Scheme

3.5 LINE PROTECTION ON TRANSMISSION CIRCUITS

3.5.1 Pilot Wire

In the pilot wire method, the phase angle of the current entering one terminal of a transmission line section is compared with the phase angle of the current leaving the other terminal. This system can only be used for line lengths of 10 miles or less because of the phase shift introduced by the pilot wire itself.

3.5.2 Phase Comparison

This method is essentially the same as the Pilot Wire method, but uses transmit and receive modems to extend the range to a few hundred miles.

3.5.3 Transfer Tripping

In this method, signals are sent between the terminals of the transmission line section being protected[7]. (Information may be sent independently in each direction.) Three forms of transfer tripping are used and are described briefly below.

a) DIRECT UNDERREACH TRANSFER TRIPPING (DUTT). This system uses underreaching relays at each terminal of the protected line section and requires simultaneous signaling in each direction. An underreach relay senses an abnormal condition over a *portion* of the protected zone.

b) PERMISSIVE UNDERREACHING TRANSFER TRIPPING (PUTT). The term 'permissive' indicates that functional cooperation between end relays is required before control action can take place. In this system, each end of the protected zone is equipped with underreach (U) and overreach (O) relays. The U relays trip local breakers for faults in the covered portion of the zone; the O relays trip the controlled breaker only if they detect a fault in conjunction with the U relay at the opposite end of the zone.

c) PERMISSIVE OVERREACHING TRANSFER TRIPPING (POTT). In the POTT system, both ends of the zone are equipped with O relays, and the breakers are tripped only when both O relays detect a fault.

The most reliable of the three systems is DUTT, but POTT has the highest security (i.e., it operates only for faults within the protected zone).

3.5.4 Directional Comparison

In this method, each end of the protected zone is equipped with two 0 relays. One of these (the blocking relay) senses the line outside the protected zone and blocks the tripping of breakers if the faults are outside the protected zone.

Loop systems that do not use communications channels use directional comparison devices, with backup, at each end of the zone. There is no coordination of breaker tripping in this case (see Figure 3-12).

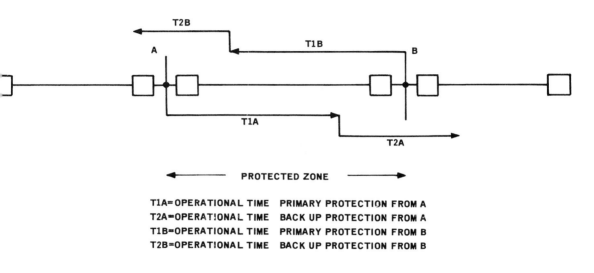

T1A=OPERATIONAL TIME PRIMARY PROTECTION FROM A
T2A=OPERATIONAL TIME BACK UP PROTECTION FROM A
T1B=OPERATIONAL TIME PRIMARY PROTECTION FROM B
T2B=OPERATIONAL TIME BACK UP PROTECTION FROM B

FIGURE 3-12

De-energization Times for a Double-Feed System
Without Communications Channel

3.5.5 Relay Types

Since the de-energization time consists mainly of the relay operation time, the selection of relays for use on protection systems is of great importance. On the majority of transmission systems, impedance relays (which respond to the voltage/current ratio) are used for the protection of lines. Impedance relays are also known as distance relays, since the impedance to which they respond is a measure of the distance.

Impedance relays provide excellent discrimination by limiting operation to a given impedance range. Modern static relays are usually all solid-state, and the absence of moving parts reduces tripping times to as low as 0.75 cycle (13 ms). Unfortunately, they are still expensive, and their application is generally limited to circuits with few branchings.

Other types of relays used in transmission lines include:

a) OVERCURRENT RELAYS. These are set to operate at a current that is sufficiently above the normal load current to give security. Their operation time is about 2 cycles.

b) TIME OVERCURRENT RELAYS. The operation time of these relays is a function of the current. For faults that occur at currents close to the normal load value, the operation time can be as high as 2 seconds. Such relays are usually used as backup on transmission circuits.

3.5.6 Line Protection on Subtransmission and Distribution Circuits

In both the Ontario Hydro and the Hydro Quebec networks, subtransmission and distribution circuits are radial. Because such circuits require many branchings, their protection necessitates almost exclusive use of interrupting devices with clearing times that depend on the values of the fault current.

Coordination in subtransmission and distribution circuits is less complex than in transmission circuits, but may still involve two or more of the following interrupting devices:

a) overcurrent relays and breakers,

b) oil or air reclosers (which have de-energization times similar to overcurrent relays and breakers),

c) fuses to protect lines or transformers, and

d) line sectionalizers, either automatic or manual. (These devices are not designed for line circuit interruption and do not, therefore, play any role in the de-energization of circuits)

Approximately 70 percent of all faults on overhead systems are of a nonpersistent nature, and thus up to four attempts may be made to reclose a line. Figures 3-13 to 3-15 show a protective arrangement, its coordination chart, and the resulting de-energization times for a distribution circuit. Because of the operating characteristics of the interrupting devices, the longest durations are those associated with faults at the end of the line, and the sum of the energization times can be greater than 3 seconds. Dead times between successive reclosing attempts are fixed with electromechanical reclosers, but are determined by preselected settings on electronically controlled reclosers.

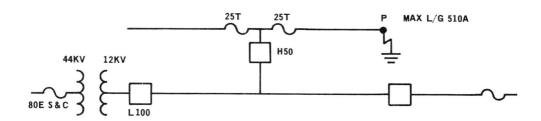

FIGURE 3-13

Example of a Protection Scheme on a
Distribution Circuit (Permanent Fault at P)

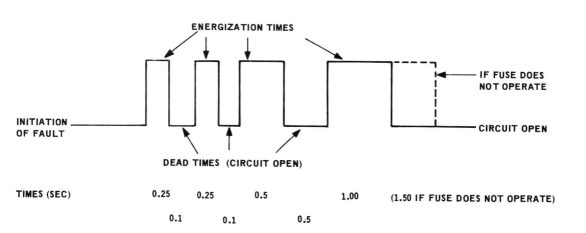

FIGURE 3-14

Resulting Times

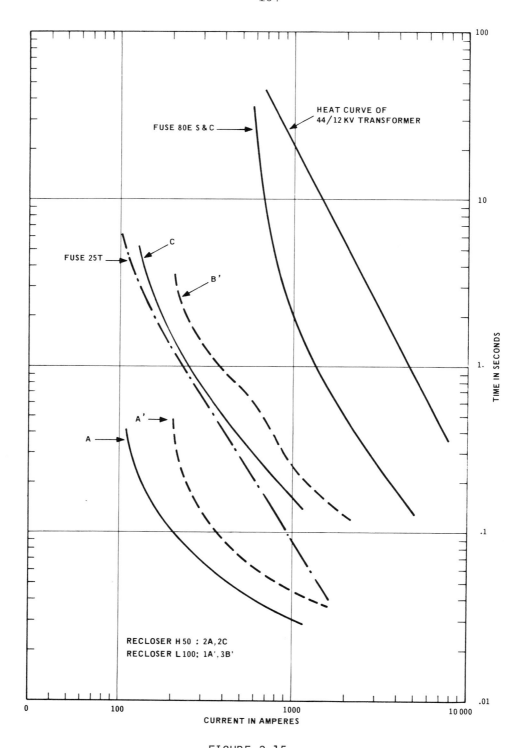

FIGURE 3-15

De-energization Times for the Protection System of Figure 3-13

fault capacity) of a power system determines the level of fault current

105

3.6 IMPEDANCE-RESISTANCE (X/R) RATIOS

The X/R ratio of a circuit in an abnormal condition determines the severity of the overshoot effect caused by the sudden connection of a voltage to an R-L circuit. The effect is characterized by increased inducing current and the saturation of ferromagnetic components.

X/R ratios vary according to the location of the fault. Generally, they have large values at the source and decrease hyperbolically with distance along the line. The X/R ratio is generally greater on transmission circuits than on distribution circuits.

3.7 THE EFFECTS OF SYSTEM MVA CAPACITY

The MVA (megavolt-ampere) capacity (or the corresponding MVA fault capacity) of a power system determines the level of fault current available at the source. However, as the distance of the fault from the source increases, the MVA capacity of the power system has a decreasing effect, until, at a few miles from the source, fault levels become virtually independent of the capacity, as shown in Figure 3-16. (This distance is typically 20 miles on a transmission line and 5 miles on a subtransmission line.) Thus, for long communications lines without protectors or other breakpoints, where the induced voltage would cause severe interference, the situation is not likely to change in future when higher capacity power systems will be involved.

However, the communication systems with short spans between break-points (such as the LD-1) will almost certainly be affected by changes in MVA capacity.

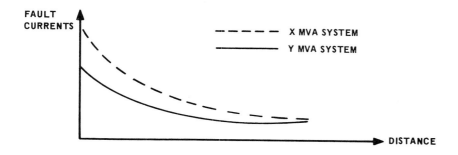

FIGURE 3-16

Variation of Fault Current Along a Line

The relationship between the growth of the capacity of a power
system and the level of interference on a communications system
varies with the relative position of the fault in the network (see
Figure 3-17). An exposure between a communications line and a
component of a power system is, in effect, a window through which
the communications line 'sees' all the power network. Even if a
fault occurs on a part of the power network outside the exposure
window, the effects of the fault will be present on the exposed
communications line with varying degrees of severity.

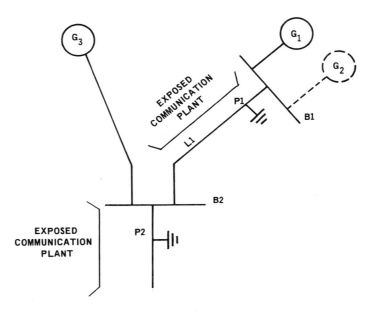

FIGURE 3-17

The Effect of Increasing the Capacity of a Power System

To determine the levels of interference on a specific line,
the model shown in Figure 3-18 can be used.

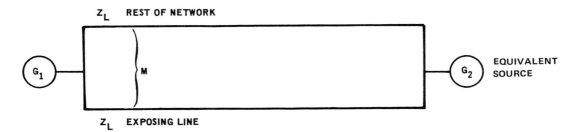

FIGURE 3-18

Model for the Prediction of Interference

3.8 THE POWER NETWORKS IN BELL CANADA TERRITORY

A summary of total electric power output during 1970 in Ontario and Quebec is provided in Table 3-3 . As indicated, in addition to the major suppliers of electrical power in Ontario and Quebec (Ontario Hydro and Hydro Quebec — see Figures 3-19 and 3-20), many other non-major companies contribute to the total power supply (see Table 3-4). However, these nonmajor suppliers produce only a small percentage of the total and this percentage is decreasing annually. Because of the relatively small contribution of nonmajor suppliers and the fact that nonmajor suppliers have small networks which do not contribute significantly to disturbances to the communications network, only Hydro Quebec and Ontario Hydro networks will be considered here.

To optimize load distribution, reliability, and continuity of service, transmission networks are interconnected at both provincial and regional levels. The transmission network of Hydro Quebec inter-connects nine of the ten regions, and the seven regions of Ontario Hydro are all interconnected. In addition, the two networks are interconnected with major U.S. networks within the Canada–United States Eastern Interconnection (CANUSE) grouping. Table 3-5 gives some basic statistics for the two Canadian networks.

TABLE 3-3

Outputs of Major and Nonmajor Suppliers
of Electric Power (1970)[8]

SOURCE	SUPPLIER	SYSTEM OUTPUT (MVA)	
		ONTARIO	QUEBEC
Thermal	Major	6 532	667
	Nonmajor	287	77
Hydro	Major	6 190	9 965
	Nonmajor	607	3 314
TOTAL		13 616	14 023
PERCENTAGE OF TOTAL PRODUCED BY MAJOR SUPPLIERS		93	76

TABLE 3-4

Nonmajor Suppliers of Electric Power in Ontario and Quebec

QUEBEC	ONTARIO
Anglo Canadian Pulp & Paper Mills	American Can
Ayers	Allied Chemical
Alcan	Atomic Energy of Canada Limited
Bellerive Veneer & Plywoods	Continental Can Company of Canada
Canadian Celanese	Canada & Dominion Sugar Co. Ltd.
Canada & Dominion Sugar Co. Ltd.	Canadian General Electric
Canadian International Paper	Canadian Niagara Power
Donnacona Paper	Cambell Ford Public Utilities Commission
Domtar Pulp and Paper	Canada Starch Co.
E.B. Eddy	Dryden Paper
Electric Reduction	E.B. Eddy
Gaspé Copper Mines	Eddy Forest Products
Gulf Power	Ford Motor
Gaspesia Pulp & Paper	Gananoque Electric Light & Water
Hart Jaune Power	Great Lakes Paper
Iron Ore	Great Lakes Power
James MacLaren	Goodyear Tire
Lorrain Mining	Huronian
Municipalité de Jonquiere	Hiram Walker & Sons
Manicouagan Power	McFadden Lumber
MacLaren Quebec	Ottawa Hydro
Noranda Mines	Ontario Minnesota Pulp & Paper
Ogilvie Flour	Ontario Paper
Ottawa Valley Power	Orillia Water Light & Power
Price Company	Polymer
Pembroke Electric Light	Pembroke Electric Light
Quebec Cartier Mining	Peterborough Hydraulic Power
Quebec North Shore Paper	Steel of Canada
Romaine Electric	Spruce Falls Power & Paper
Saguenay Power	Strathcona Paper
Smelter Power	St. Lawrence Seaway Authority
Thurso Pulp & Paper	Thunder Bay Public Utilities Commission
City of Riviere-du-Loup	Trent University
City of Sherbrooke	

TABLE 3-5

Statistics on Ontario Hydro and Hydro Quebec

STATISTIC	ONTARIO HYDRO	NOTE	HYDRO QUEBEC	NOTE
Frequency of Supply	60 Hz		60 Hz	–
Total Generated MVA	12 700	2	11 000	4
No. of Customers	2 400 000	3	1 243 000	6
Territorial Divisions	7 Regions	5	10 Regions	5
No. of Employees	23 000	3	12 200	4
Route Miles of Networks				
1.2 MV	–	–	600	8
735 kV	–	–	2020	6
500 kV	435	3	–	–
3.5 kV	–	–	1456	6
230 kV	4670	3	2536	6,7
115 kV	5170	1,3	2721	6
69 kV	170	3	1400	6
44-27 kV	6880	–	–	–
Primary Feeder Distribution	52 500	3	43 120	4

NOTES:
1. Ontario Hydro still operates a few lines of 115 kV at 25 Hz in the Niagara region and in Northern Ontario.
2. Source – Reference 8.
3. Source – Reference 9.
4. Source – Reference 10.
5. The responsibilities of the regions include construction and maintenance of subtransmission and distribution systems according to guidelines set by the head offices in Toronto and Montreal. Regions in Quebec are subdivided into Districts; regions in Ontario are subdivided into Areas, with an average of ten Areas per Region.
6. Source – Hydro Quebec, Systems Planning, P.St. Onge, December, 1972.
7. Includes 815 miles of 161 kV lines.
8. Planned.

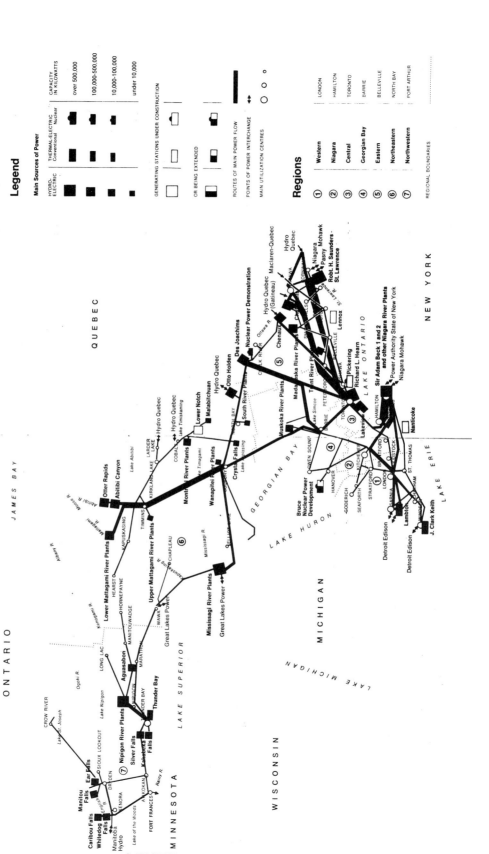

Legend

Main Sources of Power

	HYDRO-ELECTRIC	THERMAL-ELECTRIC Conventional	THERMAL-ELECTRIC Nuclear	CAPACITY IN KILOWATTS
	■	■	■	over 500,000
	■	■	■	100,000-500,000
	■	■	■	10,000-100,000
	■			under 10,000

GENERATING STATIONS UNDER CONSTRUCTION

OR BEING EXTENDED

ROUTES OF MAIN POWER FLOW

POINTS OF POWER INTERCHANGE

MAIN UTILIZATION CENTRES ○ ○ ○ ○

Regions

① Western — LONDON
② Niagara — HAMILTON
③ Central — TORONTO
④ Georgian Bay — BARRIE
⑤ Eastern — BELLEVILLE
⑥ Northeastern — NORTH BAY
⑦ Northwestern — PORT ARTHUR

REGIONAL BOUNDARIES

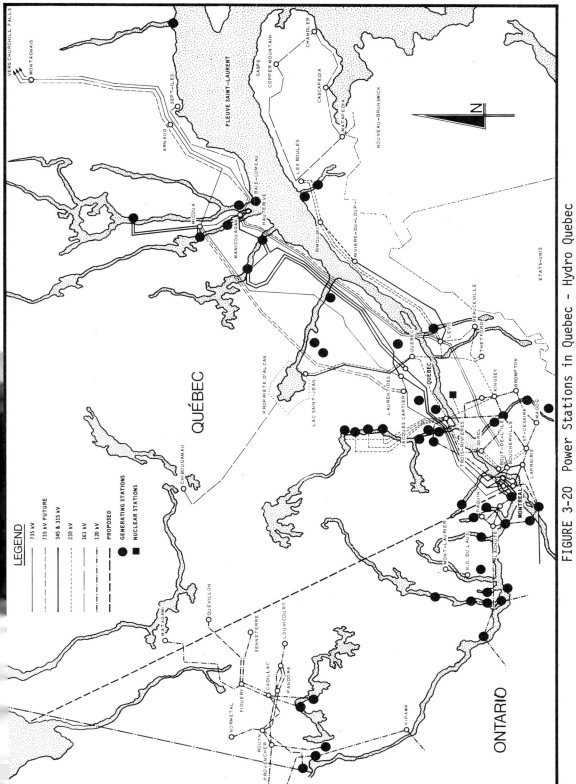

FIGURE 3-20 Power Stations in Quebec - Hydro Quebec

3.9 LOCATION OF COMMUNICATION PLANT WITH RESPECT TO THE POWER NETWORK

The outside plant components of power and telephone systems correspond in terms of circuit lengths, circuit importance, and the group of communities or customers that each class serves (see Table 3-6).

TABLE 3-6

Correspondence Between Power and Communication Outside Plant Facilities

POWER	COMMUNICATIONS
Transmission	Toll Lines
Subtransmission	Interoffice Trunks
Distribution	Exchange Distribution

Because of right-of-way constraints, the distribution lines of both telephone and power companies are usually on the same street or alley. Economic considerations and municipal regulations may force these lines to be only feet — or even inches — apart. Telephone toll lines and power transmission lines run between the same communities and must, of necessity, be on roughly parallel routes.

Although not proved statistically, there is also a close correspondence between power and telephone lines in terms of troubles. Experience has shown that power troubles on telephone toll lines are usually caused by transmission circuits, and that power troubles on telephone distribution lines are usually caused by subtransmission or power distribution circuits. Correspondence with the associated power system class is also found for the characteristics of power interference on a given class of telephone circuit.

Although the noise domain will not be dealt with here in detail, the following notes are of interest.

a) Transmission lines usually have a good balance, but may generate rf interference due to corona.

b) On distribution systems, loads and much of the line mileage are single phase. The majority of such systems outside the subtransmission network are of the grounded-neutral type.

c) Urban distribution circuits are generally short, and have high load densities. The interfering waves are contributed equally by load and transformer excitation currents.

d) Interference in rural areas is greater because:

1) Some of the plant may be unshielded or open wire.

2) Cable plant is smaller and usually has a higher resistance to ground.

3) Shields have relatively higher resistance.

4) Unlike metropolitan areas, there are no extensive underground water systems, gas systems, multiple cable runs, and other conducting objects to share earth return currents.

All of the above factors add up to reduce shielding available in rural areas as compared to metropolitan environments.

3.10 CAPACITIVE COUPLING

3.10.1 Mechanism and Formulas[36,37,38,39]

A net voltage can appear on a communication wire because of its position in an electric field generated by the voltages on a power system (see Figure 3-22).

The coupling admittance (predominantly capacitive) between the interfering source and the communication conductor is due to the electric force in the air normal to the conductor and ground. The open-circuit voltage induced appears in shunt with the interfered line; it is independent of the length of exposure and is proportional to the voltages to ground of the interfering line (see Figure 3-21).

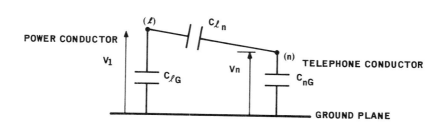

FIGURE 3-21

Capacitive Coupling Between Two Lines

With conductor (n) insulated, the open circuit voltage is approximately

$$V_n = V_1 \frac{C_{\ell n}}{C_{\ell n} + C_{nG}} ,$$ where $C_{\ell n}$ and C_{nG} are capacitances/unit length.

The open-circuit voltage is independent of the length of exposure, provided the exposure is uniform.

114

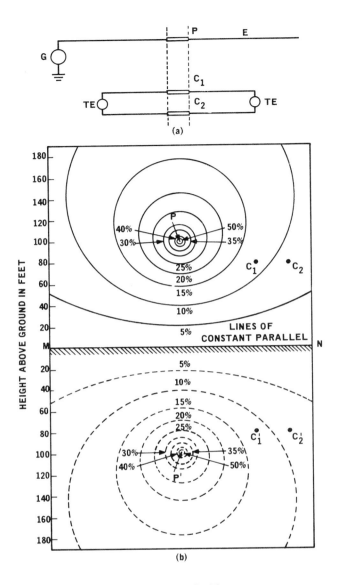

FIGURE 3-22

Electric Induction from a Single-Phase Ground Return
Power Circuit on a Communication-Circuit Conductor

a) Elementary Section
b) Equipotential Fields

For a wire that is 20 feet from the ground and 4 feet away from a power conductor (see Reference 37)

$$C_{nG} \simeq 7 \times C_{\ell n}, \quad \text{so that } V_n \simeq 0.125 \ V_1.$$

Then the current in line (n) is given by

$$J_n = V_1 \ (j\omega C_{\ell n}\ell).$$

The current is thus dependent on the length of exposure.

Generally, the determination of the voltage induced on a communication conductor through capacitive coupling with n-1 power conductors — including skywires — requires knowledge of 2n relations describing the electric conditions of the system. Of these equations, n characterize the relationships of charges to voltages between all the conductors involved. The electric constraints on the conductors provide the n other relations required to solve this system of 2n unknowns.

The equations expressing the relationships of voltages to charges are of the form:

$$V_1 = P_{11}q_1 + \cdots + P_{1n}q_n$$

$$V_n = P_{1n}q_1 + \cdots + P_{nn}q_n$$

where V_1 is the voltage-to-ground of conductor 1

(If the separation between the communication wire and a 3-phase power system is large, only the unbalanced — zero sequence voltage will have to be considered.)

q_i is the charge on conductor i

P_{ij} is the 'potential coefficient' between conductors

Considering Figure 3-23,

$$P_{ij} = 18 \times 10^9 \log_e \frac{D_{ij}'}{a_{ij}} \tag{1}$$

$$P_{jj} = 18 \times 10^9 \log_e \frac{D_{jj}'}{a_{jj}}$$

where a_{jj} is the radius of conductor (j). The other symbols are defined in Figure 3-23.

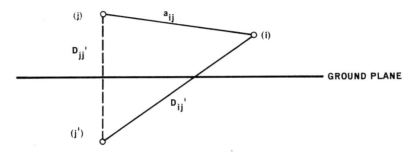

FIGURE 3-23

Definition of Symbols Used in Equation (1)

The coefficients P_{ij} have the dimension of inverse capacitance. Since both height and radius increase with increasing voltage, a typical value for a power line conductor — independent of the voltage class — is given by

$$P_n = 1/7 \ (pF/m)^{-1}, \text{ or its inverse } C_{11} = 7 \ pF/m.$$

The electrical constraints on conductors involve

a) the forced voltage applied, and

b) the relationships of the charges.

Thus, on an isolated telephone conductor, $q_n = 0$, and the equations written in conjunction with Figure 3-21 become

$$V_{no} = V_1 \frac{P_{1n}}{P_{11}} \ .$$

For a grounded conductor, $V_n = 0$,

$$q_n = -\frac{V_1}{18 \times 10^9} \cdot \frac{\log_e \frac{D_{1n}}{a_{1n}}}{\log_e \frac{D_{11}'}{a_{11}'} \cdot \log_e \frac{D_{nn}'}{a_{nn}'} - (\log_e \frac{D_{1n}}{a_{1n}})^2}$$

$$\simeq -\frac{V_1}{18 \times 10^9} \cdot \frac{\log_e \frac{D_{1n}}{a_{1n}}}{\log_e \frac{D_{11}'}{a_{11}'} \cdot \log_e \frac{D_{nn}'}{a_{nn}'}} ,$$

and the current per unit length,

$$J_n = -j_\omega \, q_n = j_\omega \, V_{no} \, C_{nn}.$$

The charging current on one wire decreases as the number of wires increases. An approximate formula for the value of the current is

$$J_n = j_\omega \, V_{no} \, C_{nn} \times \frac{3}{N + 2}$$

where N is the number of communication wires present. Cases of slanting exposures can be solved by integration.

The above formulas are derived from theories valid for electro-static fields. However, they hold very closely for the slow time-varying fields associated with power systems. Electric induction on communication wires is mitigated by drainage and shielding. For example, drainage on open wire circuits is accomplished by the use of 104A or 108A units (see Figures 3-24 and 3-25).

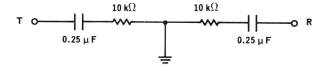

FIGURE 3-24

104A Drainage Unit

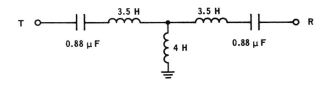

FIGURE 3-25

108A Drainage Unit

Grounded wires close to the communication wire provide shielding.
The coefficient is

$$\eta = \frac{Y_{sn} - Y_{sG}}{Y_{sn}}$$

where

Y_{sn} = admittance per unit length between shield and conductor,

Y_{sG} = admittance per unit length of shield to ground.

A wire about 1 foot from the conductor and 30 feet above ground
yields a coefficient of 0.8.

Grounded conductive enclosing shields provide absolute screening
against electric induction.

3.10.2 The Relative Significance of Electric Induction

Electric induction plays a major role in the noise interference
on older circuits such as open wire. With the effective shielding
provided by grounded sheaths on practically all the cable plant of
Bell Canada, electric induction is no longer a problem even from a
noise point of view. There is a possibility of mild shock to
personnel working on aerial wires and cables. However, because of
the large impedance of the source being derived from the capacitance
between conductors and ground and that of the paths involved, the
resulting current is small, thus making shock by electric induction
a hazard only because of its 'surprise' effect.

Personnel working near high voltage lines (above 100kV)
may become charged by the electric field produced by such lines.
If a grounded object such as an aerial cable strand is then contacted,
a discharge to the strand may occur, and the resulting sensation of
shock may give the impression that the strand is continuously
energized.

3.11 RESISTIVE COUPLING

3.11.1 Mechanism and Formulas

General Resistive Coupling can be divided into two categories (Figure 3-26):

a) contact between the communication facility and an energized wire (with or without contact impedance), and

b) paths through the soil between grounds on power and telephone systems and via bonds between telephone plant and power multigrounded neutral.

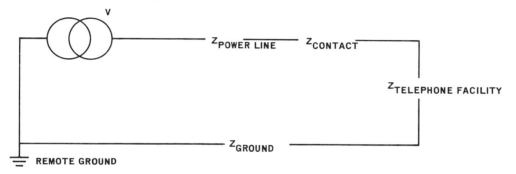

FIGURE 3-26

Power Contact (Equivalent Single-Line Representation)

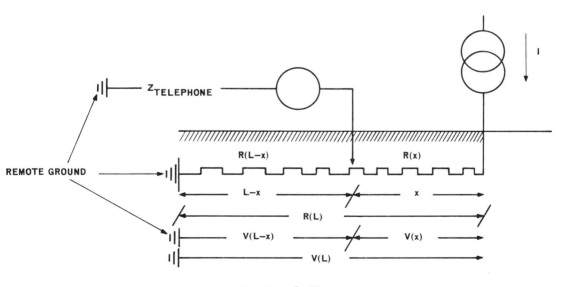

FIGURE 3-27

Ground Potential Rise (Distance Representation)

The total voltage V(L) developed at the point of fault is given
by the product of the fault current and the impedance to remote ground
of the grounding structure (see Figure 3-27). The distribution of
voltage to remote ground usually has the hyperbolic form $1/(A + Bx)$,
where A and B are constants and x is the distance from the source in
miles. Consequently, the following relationships can be made:

$$V(L) = I \cdot R(L)$$

$$V(L-x) = I \cdot R(L) \left[1 - \frac{1}{1 + R(L) \cdot B \cdot x} \right]$$

where B has the dimension of admittance per unit length.

Although a detailed study of ground potential rise (GPR) is made
in Section 5 of this report, it is useful at this point to note
the combination of interference from resistive and magnetic
couplings.

A telephone facility might experience simultaneous couplings
from both resistive and magnetic coupling. Such is the case for
communication services to power stations or for joint-use situations.

For GPR to become a problem, the connection to ground of the
telephone facility has to be close enough to the source. With
existing fault current levels, earth resistivities, and power
grounding arrangements, a connection to ground over 500 feet from
the source does not represent a threat to the telephone facility.
Under this condition of proximity and with the usually low levels
of earth resistivity in the Bell Canada territory, the coupling
impedance from magnetic induction is mostly reactive, e.g., with
100 feet separation and in an earth resistivity environment of
$100 \ \Omega m$, the X component is $0.035 \ \Omega/kft$, and the R component is
$0.007 \ \Omega/kft$ (see Reference 11). It is therefore more representative
to combine magnetic and resistive coupling on a root-sum-square
basis instead of by arithmetic addition.

3.11.2 Power Contacts

Potential power contacts exist at aerial crossings, joint-use
constructions, and in random burial situations. In buried or underground arrangements, however, experience has shown that the occurrence
of direct contacts is so small that it is not considered a significant
hazard.

Acceptance criteria for aerial joint use and crossings include
both structural and electrical requirements. In Canada, the structural
requirements are outlined in CSA Standard C 22.3[12]. References
readily available to plant personnel are the AG17 series and
Division G20 of the 9-digit series. CSA Standard C 22.3 also gives
horizontal and omni-directional clearance limits under maximum sag

condition. However, even if all the structural requirements were met, there may be extreme or chance situations that lead to power contacts, e.g., cars knocking down pole lines, foreign objects falling on the lines, and very severe sleet storms. Structural protection against all possible causes of contact is economically prohibitive. Because contacts are unavoidable, their effect must be controlled through proper electrical coordination with power systems.

3.11.3 Electrical Requirements

The safety of the public and of personnel, the minimization of public property damage, and the limitation of plant damage form the basic objective of aerial JU design.

Experience has shown that, for contacts other than local breakdowns, most of the damages are due to thermal effects.

To meet the stated objectives, the only recommendation given in the References (13, 14, 15) concerning the design of JU is that prompt de-energization of the power circuit must be ensured. In those references, a comparative description of various power systems is given based on the speed of clearing a fault. Delta or Unigrounded-Y circuits are not recommended for JU.

To implement the indefinite recommendations of References 13, 14, 15, operating engineers in Bell Canada have adopted a set of limits for JU with lines up to 20 kV phase-to-phase voltage. These limits seem to have gained acceptance within the power industry.

Given that the bonding and grounding arrangements are satisfactory,

a) the total de-energization time must be less than or equal to 3 seconds, and

b) damages on the telephone plant must be limited to one span.

To adapt electrical requirements to existing aerial plant (ALPETH-PIC), and to provide limits for use with voltages in the range 20 to 69 kV, the following criteria were recently proposed[40]:

1) total de-energization time, $\leq$ 3 seconds,

2) total voltage drop between point of contact and ground, $\leq$ 3000 V_{rms},

3) I^2t shall not exceed the no damage limit of the cable sheath.

Also to ensure that the above limits will be met, a more frequent (every 500 feet) bonding to the power neutral was recommended.

These limits are now under discussion with the power authorities. Without attempting to comment on the validity of the limits, the following remarks can be made:

- Limit (1), $\leq$ 3 seconds: In most multigrounded distribution and subtransmission systems, the total duration of a fault is usually less than 3 seconds, even for faults at the end of the line. Faults involving part of the communication plant in the return path would have lower fault resistances and, therefore, shorter durations.

- Limit (2), $\leq$ 3000 V_{rms}: A large percentage of aerial plant is of the ALPETH-PIC type. Dielectric strength between the conductor and the sheath of such cables is provided by the polyethylene insulation of the conductor and a MYLAR tape surrounding the core. For a 26-gauge cable, the strength is 5 kV dc applied for 3 seconds. Although there is no known relationship between ac and dc breakdown, it is estimated that 60 Hz core-to-sheath breakdown of this cable would occur at 2500 V_{rms} in 3 seconds[16]. Therefore, if the voltage developed between the point of contact on the sheath and a remote ground is less than 3000 V_{rms}, it is likely that breakdown, if it occurs, will involve only peripheral conductors. For example, if n peripheral conductors share the current with the sheath, the impedance of the telephone plant that determines the drop will decrease, but the increase in fault current will not be proportional because of the masking effect of the power line impedance.

- Limit (3) is usually not reached except in a direct contact to drop because conductors in terminal stub will fuse open before 'c' drop wire.

Limits (1) and (2) are independent. Joint use is not recommended if either (1) or (2) is not met. Because of this characteristic, there seems to be some redundancy in the specification of an I^2t limit for the cable sheath. For example, assume that the lowest sheath impedance from the point of contact to the nearest ground connection is Zs, and that the lowest ground resistance connection is R_g. The maximum I^2t is given by

$$3\left[\frac{3000}{|Zs + R_g|}\right]^2.$$

A model representing a typical case can be established, and the I^2t limits can be derived from (1) and (2).

Although mentioned in Design Memorandum 71-02, recommendations on I^2t cannot be fully implemented at present, since I^2t characteristics for various types of cables have not yet been fully documented. I^2t values for strands are quoted in Design Memorandum 71-02 on the assumption that the contact will always occur between the live wires and the strand. Of course, it will be useful to know the combined I^2t limit for strand and shield, which will be a more representative figure taking into account the actual physical construction of plant.

3.11.4 Incidences of Power Contact in Bell Canada Plant

Approximately 60 percent of Bell Canada aerial distribution plant is in joint use[17]. After reviewing the data supplied by four Bell Canada areas (Western, Toronto, Central, and Eastern) covering periods between 1968 and 1972, it appears that, whatever the criteria of acceptance, the construction of plant and the care taken in coordination with power systems are excellent (see Table 3-7).

Of the 154 contacts reported for the period 1971-72, 71% occurred on joint-use pole lines, 21% occurred at span or pole crossover locations, and the remaining 8% were not specified. The causes of these contacts are shown in Table 3-8. The average trouble rate is 4 cases of contact per year per kilomile of aerial cable.

TABLE 3-7

Relative Damage in Bell Canada Plant

TYPE OF DAMAGE	TOTAL % OF INCIDENTS
Sheath Burns	83
Strand Burns	12
Pair Fused, Open or Crossed	55
Carbons Replaced	7
Other (terminals, drop wire etc.)	12

TABLE 3-8

Causes of Contacts in Bell Canada Plant

CAUSE	TOTAL % OF INCIDENTS
Weather	26
Trees	9
Man (cars, trucks, high loads, etc.)	13
Unknown and Miscellaneous	52

Table 3-9 shows a complete breakdown of the cases and the extent of the damage. Due to the lack of a definite pattern and the size of the sample, it is not possible to draw meaningful conclusions with regard to causes or the position of the contacts in the telephone plant.

Because a greater percentage of the telephone plant will in future be buried (the best current estimate is 90%), and because of the small rate of contacts on existing plant, there appears to be no justification at present for further statistical characterization of the subject. This does not, however, preclude the need for technical studies on joint use with other voltage classes. For instance, joint use with voltages up to 150 kV ac is now being considered by Bell Canada. Although of better structural design and with faster and more reliable de-energization schemes, such lines could generate very high currents if contacts occur near the source of supply. However, at these high voltages the protection of linemen from shocks caused by electric induction might become the controlling factor in joint-use design.

The criteria of acceptance for new systems will have to ensure that damages and their frequency of occurrence are the same or less than the present ones.

3.12   MAGNETIC COUPLING

3.12.1 General

In terms of frequency of occurrence, magnetic coupling is the most important of the three modes of coupling. However, although the mechanisms of this inductive interference were studied in detail during the thirties, the statistical aspect of the problem, which is important from a protection standpoint, was largely neglected perhaps because of a concentration of interest on lightning. Consequently, few statistics were collected on frequency of occurrence, location, and severity of inductive interference. Recent studies[19,20]

TABLE 3-9

Power Contacts in the Central, Eastern, Western, and Toronto Areas

AREA	YEAR	NUMBER OF INCIDENTS	EXTENT OF DAMAGE: INCIDENT INVOLVING					CAUSES OF CONTACT					POWER				LOCATION OF CONTACT CROSSOVER			
			SHEATH	PAIRS	STRAND	CARBONS	OTHER	WEATHER	TREES	MAN	IMPROPER DESIGN	OTHER	SEC	DTS 2.0-15.9	SUB TRN	JU	SPAN	POLE	RED	OTHER
Central	1968	21	16	11	1	1	6	6	4	4	2	7	1	13	2	9	0	3	0	18
	1969	15	14	10	2	1	1	3	2	3	0	7	2	7	0	9	4	1	1	9
	1970	23	17	12	2	2	3	0	7	0	1	16	3	3	0	19	0	3	1	19
	1971	16	8	6	2	3	2	0	5	3	1	8	1	6	0	14	2	0	0	14
	1972	10	6	6	0	5	7	1	2	2	0	5	0	5	2	5	5	0	0	5
	TOTAL (68–72)	85	61	45	7	13	13	10	20	12	4	43	7	34	4	56	11	7	2	65
	TOTAL (71–72)	26	14	12	2	8	9	1	7	5	1	13	1	11	2	19	7	7	2	19
Eastern	1970	8	7	5	1	0	0	2	0	0	6	0	1	0	0	2	2	1	1	4
	1971	12	10	7	5	1	0	4	0	4	4	0	2	5	1	5	1	4	0	7
	1972	33	33	15	3	0	0	17	0	1	15	0	5	0	0	30	0	1	0	32
	TOTAL (68–72)	53	50	27	9	1	0	23	0	5	25	0	8	5	1	37	3	6	1	43
	TOTAL (71–72)	45	43	22	8	1	0	21	0	5	19	0	7	5	1	35	1	5	1	39
Western	1971	10	7	5	5	1	2	0	3	3	4	1	2	3	0	7	1	0	1	8
	1972	32	31	18		0	1	14	3	3	12	0	1	6	0	26	2	3	0	27
	TOTAL (71–72)	42	38	23	5	1	3	14	6	6	16	0	3	9	0	33	3	3	1	35
Toronto	1971	26	19	16	2	0	3	0	0	1	25	5	3	2	0	15	9	2	5	10
	1972	15	14	11	1	1	4	4	0	3	8	3	1	2	0	8	3	0	1	11
	TOTAL (71–72)	41	33	27	3	1	7	4	0	4	33	8	4	4	0	23	12	2	6	21
TOTAL	71–72	154	128	84	18	11	19	40	13	20	69	21	15	29	3	110	23	17	10	114

outside North America now indicate a growing concern for the
characterization of this type of interference because of increases
in power capacities, closer proximities between telecommunication
and power networks, and more sensitive telephone plant.

In the following pages, some statistical information on the
existing magnetic coupling situation in Bell Canada territory is
provided. The data was collected from existing information because of the difficulties involved and the extent of monitoring
required to produce a new statistical description.

3.12.2 Mechanisms and Formulas

Consider the circuit shown in Figure 3-28. The magnetic
fields created by the time-varying currents circulating in a and a'
induce longitudinal and opposite electric forces on wire b. If the
distances ba and ba' were equal, the net emf would be nil. However,
this condition of symmetry is never met in inductive exposures
because the return path is through earth for most of the fault
conditions and the power circuit is seldom symmetrical with respect
to the communication network.

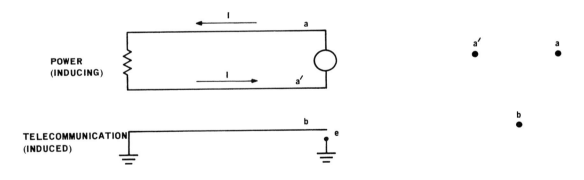

FIGURE 3-28

Magnetic Induction in Telecommunication Lines

The current distribution in the ground is such that heat
loss and magnetic energy are minimal. Minimization of heat loss
requires the circulation of current through a large volume of earth
(the greater the resistivity, the larger the volume). Minimization
of magnetic energy for time-varying currents requires concentration
of opposite line currents. These effects result in a distortion of
the line currents and a penetration of the equivalent return current
to a depth inversely proportional to frequency and directly proportional to resistivity.

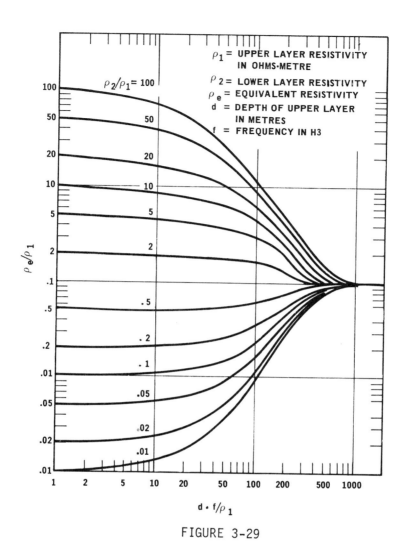

FIGURE 3-29

Equivalent Earth Resistivity for Two-Layer Earth Structures
(After Reference 21)

The evaluation of the coupling impedance between circuits with earth return cannot be done by standard formulas (e.g., Neumann's) which assume uniform field distributions. The first attempts to develop the correct formulas were made in the 19th century although the results obtained were not accurate in practice. Carefully conducted experiments later confirmed formulas developed independently and almost simultaneously by Haberland, Pollaczeck, Carson, and Buckholz.

These formulas were derived from Maxwell's general equations using the following assumptions:

1) Relative magnetic permeability of soil is unity.

2) Displacement currents are neglected with respect to conduction currents. (This assumption limits the applicability of the formulas up to about 60 kHz.)

3) The earth is homogeneous and of finite resistivity.

4) The length of the interfering conductor is infinite.

5) To account for the nonhomogeneity of the soil, an equivalent larger volume soil must be assumed as shown in Figure 3-29.

The main formulas for mutual inductance M in henries per meter are listed below. Other formulas have been developed to account for the finiteness of the interfering line; however, this additional sophistication does not normally result in appreciably improved accuracy.

Carson:

$$M = \frac{j\mu_0}{\pi} \int_0^\infty (\sqrt{u^2+j} - u) \, \exp(-bau) \, \exp(ca\sqrt{u^2+j}) \, \cos(aau) du$$

Pollaczek:

$$M = \frac{j\mu_0}{2\pi a^2} \left[\int_{-\infty}^0 \exp\{[u(ja+b) + c\sqrt{u^2+ja^2}]\} \, (\sqrt{u^2+ja^2} + u) du \right.$$
$$\left. + \int_0^\infty \exp\{[u(ja-b) + c\sqrt{u^2+ja^2}]\} \, (\sqrt{u^2+ja^2} - u) du \right]$$

Haberland:

$$M = \frac{\mu_o}{2\pi} \log \frac{a}{d} + \frac{j\mu_o}{\pi} \int_0^\infty (\sqrt{u^2+j} - u) \exp(-u^\alpha(b+c)) \cos(a^\alpha u) du$$

Posdnjykov [22]:

$$M = \frac{\mu_o}{2\pi} \int_0^\theta \exp(-\beta) \left\{ \left[\frac{\sin\beta - \cos\beta}{\sqrt{2\alpha d}} + \frac{\cos u \sin\beta}{(\alpha d)^2} \right] \right.$$

$$+ j \left[\exp(-\beta) \cdot \frac{\sin\beta + \cos\beta}{\sqrt{2\alpha d}} + \frac{\cos u}{(\alpha d)^2} (\exp(-\beta)\cos\beta - 1) \right] \Big\} du$$

where

μ_o = permeability of free space = $4\pi \times 10^{-7}$
u = variable of integration
σ = conductivity of ground (mhos/meter)
ω = $2\pi f$ (f in Hz)
α = $(\mu_o + \sigma\omega)^{\frac{1}{2}}$
b = height of inducing line w.r.t. ground level (metres)
c = height of induced line w.r.t. ground level (metres)
a = horizontal separation between inducing and induced line (meters)
d = length of the inducing line (metres)
θ = arctan ℓ/d
β = $\alpha d/\sqrt{2}$ (cos u)

Within the range of resistivities, frequencies, and spacings of most of the inductive exposures encountered, the results obtained with any of the above four formulas do not differ appreciably from each other. For this study, the algorithms developed by Carson were used.

All the above formulas are established for strictly parallel exposures. Oblique exposures are dealt with in the following manner. The slanted inducing (or inducted) line is divided into elementary parallel and perpendicular sections. Only the parallel sections contribute to the value of the mutual. The calculations can be made by either summing up the contributions of the elementary sections, or integrating over the range of separation (see Figure 3-30).

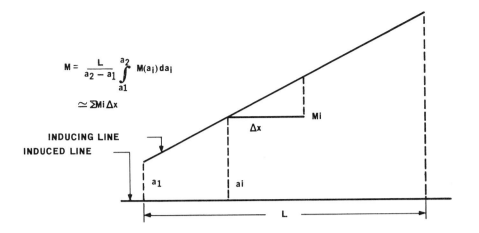

FIGURE 3-30

Equivalent Combinations of the Elementary Contributions
in the Calculation of the Resultant Mutual Inductance

3.13   THE FAULT CURRENT

3.13.1 Analysis

A fault on a power system, i.e., an unbalance condition, can
be readily analysed using the symmetrical components theory. This
theory, which is derived from the superposition theorem, states that
any unbalanced system of phasors, A_i,...., An, can be decomposed
into n balanced systems of order $0,1,2,...,n-1$. That is, the original
n phasors lead to the system $B_0,...,B_{n-1}$, where $\{B_j\} = T\{A_i\}$[23].* Thus

$$B_0 = \frac{1}{n} \sum_{i=1}^{n} A_i \quad ,$$

*The theory of symmetrical components was developed by C.L. Fortesque,
 see Reference 23.

$$B_1 = \frac{1}{n} \sum_{i=1}^{n} a^{(i-1)} A_i,$$

$$B_{n-1} = \frac{1}{n} \sum_{i=1}^{n} a^{(n-1)\,(i-1)} A_i,$$

where a is the operator $\exp[j(2\pi/n)]$.

Conversely, the system $\{A_i\}$ could be expressed as a function of the system $\{B_j\}$; i.e., $\{A_i\} = \{B_j\}$.

In the most common system, i.e., the three phase, the transformation T leads to three balanced systems. Considering the unbalanced phase to ground voltages Va,Vb,Vc, at the point of fault, the following results are obtained (see Figure 3-31):

(Order Zero) Zero Sequence: $V_0 = V_a + V_b + V_c$

(Order One) Positive Sequence: $V_1 = V_a + aV_b + a^2V_c$

(Order Two) Negative Sequence: $V_2 = V_a + a^2V_b + aV_c$

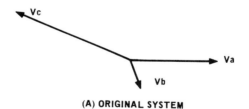

(A) ORIGINAL SYSTEM

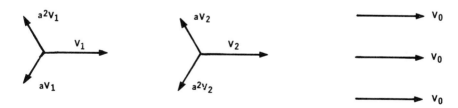

POSITIVE SEQUENCE   NEGATIVE SEQUENCE   ZERO SEQUENCE

(B) BALANCED SYSTEMS

FIGURE 3-31

Three Phase T Transformation

The net benefit of this representation in terms of magnetic
interference on communication facilities is that the only faults
that can affect telephone lines are those where zero sequence currents
can circulate. This is true provided there is no differential
coupling between communication and power facilities.

The symmetrical systems obtained through the transformation
T correspond to the following physical realities:

a)   Each of the sequence components is measurable through
     appropriate filtering.

b)   The positive sequence generates a field which creates the
     driving forces of motors.

c)   The negative sequence generates an inverse rotating field
     which actually dampens the driving forces.

For example[24], in an alternator, the rotating field generated by
the positive sequence is given by $E = k\cos(p\theta - \omega t)$, while the negative
sequence yields $E = k\cos(p\theta + \omega t)$, where

$$2p = \text{total number of poles}$$
$$\theta = \text{angular motion of the field}$$
$$k = \text{constant}$$

The types of fault which can affect communication facilities
are shown in Table 3-10.

3.13.2 Residual Current — Variation With Distance

Only phase-to-ground faults, which have both the greatest
frequency of occurrence and the greatest fault value[25] as far as
magnetic interference is concerned, will be considered because of
their relative importance.

SINGLE FEED — SINGLE CIRCUIT.   This case occurs in distribution
and most subtransmission lines. A typical circuit is shown in
Figure 3-32.

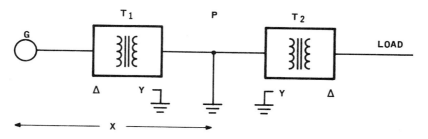

FIGURE 3-32

Single Feed - Single Circuit

Practically, the positive and negative sequences of the load are much greater than any of the reactances of the line or apparatus. Therefore, for those sequences, the circuits to the right of point P in Figure 3-32 can be considered open. Also, even if the low voltage side of T_2 was Y-connected with grounding and the load was also grounded, the zero sequence component of the load is such that as far as the zero sequence circuit is concerned, the circuit to the right of T_2 can be considered open. With the previous assumptions we have

Z 1st positive sequence $\simeq$ Z 2nd positive sequence.

The total fault current is then given by

$$I_p = \frac{3V}{Z_{1_{\omega p}} + Z_{2_{\omega p}} + Z_{o_{\omega p}} + 3R} ,$$

$$Z_1 = j[X_{1G} + X_{1T1} + X_{1\ell} \cdot x],$$

$$Z_2 = j[X_{2G} + X_{2T1} + X_{1\ell} \cdot x],$$

$$Z_o = j[(X_{oT1} + X_{c\ell} \cdot x)(X_{o\ell}(L-x) + X_{oT2})] / (X_{cT1} + X_{cT2} + X_{o\ell} \cdot L).$$

For a large X_{0T2} this reduces to

$$Z_o = X_{oT1} + X_{o\ell} x.$$

TABLE 3-10
POWER SYSTEMS FAULTS
INDUCING VOLTAGES ON COMMUNICATION FACILITIES

FAULT CONNECTION	SCHEMATIC *Only One Phase Represented	SEQUENCES — NETWORKS	TOTAL RESIDUAL CURRENT MAGNITUDE — EXACT FORM	TOTAL RESIDUAL CURRENT MAGNITUDE — SIMPLIFIED FORM (CONSERVATIVE)	SEE NOTES
Line to Ground Grounded System		1st Positive Sequence / 2nd Positive Sequence / Negative Sequence / Zero Sequence	$\left\|I_{F_1}\right\| = \left\|\dfrac{3Vn}{Z_0+Z_1+Z_2+3R}\right\|\left\|\dfrac{Z_1}{Z_1'}\right\|$	$\left\|I_{F_1}\right\| = \left\|\dfrac{3Vn}{Z_0+Z_1+Z_2+3R}\right\|$	1 to 6
Line to Ground Isolated System		Same as above for the 1st Positive sequence, 2nd Positive sequence, and Negative sequence. The impedance seen at the fault point in the Zero sequence network is the total capacitance to ground of the line.	$\left\|I_{F_2}\right\| = \left\|\dfrac{3Vn}{(-j/C\omega)+Z_1+Z_2+3R}\right\|\left\|\dfrac{Z_1}{Z_1'}\right\|$	$\left\|I_{F_2}\right\| = 3Vn\,C\omega$	7,8

FAULT CONNECTION	SCHEMATIC	SEQUENCES — NETWORKS	TOTAL RESIDUAL CURRENT MAGNITUDE — EXACT FORM	TOTAL RESIDUAL CURRENT MAGNITUDE — SIMPLIFIED FORM (CONSERVATIVE)	NOTES #
Double Line to Ground Grounded System		Same sequence division as in case of L/G Grounded Systems.	$$\left|I_{F_3}\right| = \left|\frac{(R_b + a^2 R_a - a Z_2) 3 V_n}{[Z_0 Z_1 + Z_0 Z_2 + Z_1 Z_2 + (Z_0 + Z_1 + Z_2 + 3 R_{ab})(R_a + R_b) + 3 R_{ab}(Z_1 + Z_2) + 3 R_a R_b]}\right| \left|\frac{Z_1}{Z_1}\right|$$	$$\left|I_{F_3}\right| = \left|\frac{3 V_n}{Z_0 \frac{Z_1}{Z_2} + Z_0 + Z_1 + \frac{3R(Z_1+Z_2)}{Z_2}}\right|$$	9, 10, 11
Double Fault Isolated Systems		1st Positive sequence in Original nonfaulted system 2nd Positive sequence with Positive EMFs in Branch B 3rd Positive sequence with Positive EMFs in Branch A 1st Negative sequence with Negative EMFs in Branch B 2nd Negative sequence with Negative EMFs in Branch A 1st Zero sequence with Zero EMFs in Branch B 2nd Zero sequence with Zero EMFs in Branch A	$$\left|I_{F_4}\right| = 3 V_n \left|\frac{M}{N}\right|$$ $$M = \frac{Z_{a1}''}{Z_{a1} Z_{a1}} \cdot \frac{Z_{a1}(Z_{a1}Z_{a1}'' + Z_{a1}Z_{a1}') - a^2 Z_{a1}'(Z_{a1}Z_{a1}' + Z_{a1}'Z_{a1}'')}{Z_{a1}^2 - Z_{a1}'Z_{a1}}$$ $$N = 3(R_A R_B) + Z_{a0}' + \frac{Z_{a1}''(Z_{a1}'Z_{a1}' + Z_{a1}Z_{a1}' + Z_{a1}'Z_{a1}'')}{Z_{a1}^2 - Z_{a1}'Z_{a1}} + \frac{Z_{a1}''(Z_{a2}Z_{a2}'' + Z_{a2}Z_{a2}' + Z_{a2}'Z_{a2})}{Z_{a2}^2 - Z_{a1}'Z_{a1}} + Z_{a0}Z_{a2}$$	$$\left|I_{F_4}\right| = \left|\frac{3\sqrt{3} V_n}{3(R_A + R_B)'' Z_{c0}' + Z_1 + Z_2}\right|$$	12 to 18

NOTES:

1. V_n is the *Generated* Neutral to Phase voltage.

2. Z_0 is the Zero sequence impedance seen from the fault point.

3. Z_1 is the Positive sequence impedance seen from the fault point for the 2nd Positive sequence.

4. Z_2 is the Negative sequence impedance seen from the fault point for the 2nd Positive sequence.

5. Z_1' is the Positive sequence impedance for the 1st Positive sequence $Z_1' > Z_1$.

6. Percentage L/G type w.r.t. all types of fault = 70% (W p. 358) (includes type 2).

7. C is the total Capacitance to Ground seen from the fault point. $I_{F_2} \approx 0.005$ A/kV/Mi (Radial feed).

8. The resulting fault current is negligible in ALL cases (Bell EEI Engineering Report #27).

9. The definitions of Z_0, Z_1, Z_2 are similar to those given for type 1.

10. I_{F_3}/I_{F_1} with R=0 yields $I_{F_3}/I_{F_1} \approx \frac{Z_0 + 2Z_1}{2Z_0 + Z_1}$; Z_0 is usually the controlling element $\therefore$ $I_{F_3} < I_{F_1}$.

11. Percentage of DL/G w.r.t. all types of faults = 15% (Westinghouse p. 358) (included L/L).

12. Z_{a1}, Z_{a1}', are the 1st and 2nd Positive sequence impedances seen from B.

13. Z_{a1}, Z_{a1}', are the 1st and 3rd Positive sequence impedances seen from A.

14. Z_{a2}, Z_{a2}', are the 1st and 2nd Negative sequence impedances seen from B, and A respectively.

15. Z_{a0} is the Zero sequence impedance of the path(s) connecting A to B.

16. Z_{a1} is the transfer impedance from B to A (or A to B) in either the 2nd or 3rd Positive sequence network.

17. Z_{a2} is the transfer impedance from B to A (or A to B) in either the 1st or 2nd Negative sequence network.

18. Such types of fault occur usually as a degeneration of a L/G fault on Δ systems. The percentage of such faults is 10% of all types of faults (Westinghouse p. 358).

In that case

$$I_p = \frac{V}{Ax + B} \quad ,$$

with
$$A = (2X_o\ell + X_o\ell)/3 \quad ,$$

$$B = [j(X_{1G} + X_{2G} + X_{1T1} + X_{2T1}) + 3R]/3.$$

For zero fault resistance, or if the fault is far enough from the generating point,

$$|I_p| \simeq \frac{V}{Ax + B'} \quad ,$$

with
$$B' = B - R. [25]$$

NOTE: Double line to ground faults at different points on isolated systems could yield higher values of fault currents. Such a fault occurs as the degeneration of a single line to ground fault. That is, once phase 'a' is at ground, the voltage given by $\sqrt{3} \times$ Nominal voltage stresses the sound phases which would breakdown at a point 'b' close to the original L/G fault point 'a'. The return path ab is therefore likely to be short making the total coupling impedance insignificant. Moreover, because of the insignificant number of completely isolated lines, consideration of such faults is not important.

DOUBLE FEED — SINGLE CIRCUIT. This case occurs in the inter-connected network. A typical circuit is shown in Figure 3-33.

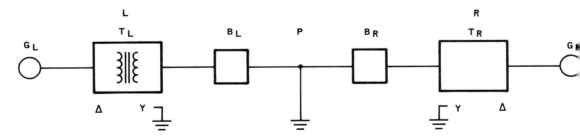

FIGURE 3-33

Double Feed — Single Circuit

The sequence of breaker operations is usually as follows:

a) Fault occurs at P.
b) Fault detected at R (or L).

3) Breaker R (B_R) opens and signal sent to L to open breaker L (B_L).

4) B_L open.

The time delay between the opening of B_R and B_L is small (approximately 2 to 4 cycles).

This sequence indicates that the fault should be analysed for the following two situations.

Case (a): Both G_R and G_L feed in the fault.

Case (b): Only G_L (or G_R) feeds in the fault.

Note: G_R and G_L are voltage sources and not necessarily generators.

Case (a)

(1) Total Fault Current*:

Impedances are given by the following formulas:

$$Z_o = \frac{[X_{TL_o} + xX_o]\ [X_{TR_o} + (L-x)X_o]}{X_{TL_o} + X_{TR_o} + LX_o},$$

$$Z_1 = \frac{[X_{GL_1} + X_{TL_1} + xX_1]\ [X_{GR_1} + X_{TR_1} + (L-x)X_1]}{X_{GL_1} + X_{TL_1} + X_{GR_1} + X_{TR_1} + LX_1},$$

$$Z_2 = \frac{[X_{GL_2} + X_{TL_1} + xX_1]\ [X_{GR_2} + X_{TR_1} + (L-x)X_1]}{X_{GL_2} + X_{GR_2} + X_{TL_1} + X_{TR_1} + LX_1},$$

where

X_{GL_1} = % positive sequence impedance of G_L**

---

* It is assumed that all impedances have been expressed in % relative to the same kVA basis (For a definition of percent impedance of line and apparatus, see Westinghouse Reference Book, pages 163, 294, 295).

** The positive sequence impedance of generators varies with time. That is, at the beginning of the fault the impedance of G (Reactance) has a value X subtransient which gradually increases to X transient and X synchronous. We have $X_{sub} < X_{Tr} < X_{sync}$. In a later section, it will be shown how to analyse the reactance variations.

X_{TL_1} = % positive sequence impedance of T_L = X_{TL2}

X_1 = % positive sequence impedance per unit length of the line = X_2

X_{GR_1} = % positive sequence impedance of G_R

X_{TR_1} = % positive sequence impedance of T_R = X_{TR2}

X_{GL2} = % negative sequence impedance of G_L $\neq$ X_{GL_1}

X_{GR_2} = % negative sequence impedance of G_R

X_{TL_0} = % zero sequence impedance of T_L

X_0 = % zero sequence impedance per unit length of the line

X_{TR_0} = % zero sequence impedance per unit of T_R

I_{Total}(Fault)=

$$\frac{1}{\sqrt{3}} \quad \frac{Base\ kVA}{Base\ kV} \times \frac{100}{Z_1+Z_2+Z_0+3R} = \frac{K'}{Z_0+Z_1+Z_2+3R}$$

Let A' =

$$\frac{-X_1^2}{X_{GL_1}+X_{TL_1}+X_{TR_1}+LX_1} - \frac{X_1^2}{X_{GL_2}+X_{GR_2}+X_{TL_1}+X_{TR_1}+LX_1} - \frac{X_0^2}{X_{TL_0}+X_{TR_0}+LX_0} .$$

Let B' =

$$\frac{X_1(X_{GR_1}+X_{TR_1}+LX_1-X_{GL_1}-X_{TL_1})}{X_{GL_1}+X_{TL_1}+X_{GR_1}+X_{TR_1}+LX_1} + \frac{X_1(X_{GR_2}+X_{TR_1}+LX_1-X_{GL_2}-X_{TL_1})}{X_{GL_2}+X_{GR_2}+X_{TL_1}+X_{TR_1}+LX_1} + \frac{X_0(X_{TR_0}+LX_0-X_{TL_0})}{X_{TL_0}+X_{TR_0}+LX_0}$$

Let C' =

$$\frac{(X_{GL_1}+X_{TL_1})(X_{GR_1}+X_{TR_1}+LX_1)}{(X_{GL_1}+X_{TL_1}+X_{GR_1}+X_{TR_1}+LX1)} + \frac{(X_{GL_2}+X_{TL_1})(X_{GR_2}+X_{TR_1}+LX_1)}{(X_{GL_2}+X_{GR_2}+X_{TL_1}+X_{TR_1}+LX1)} + \frac{(X_{TL_0}(X_{TR_0}+LX_0)}{(X_{TL_0}+X_{TR_0}+LX0)} .$$

The following expression for total fault current (case a) is now obtained:

$$I_{Total\ Fault_a} = \frac{K'}{A'x^2+B'x+C'+3R} = \frac{K}{Ax^2+Bx+C} \quad . \quad (2)$$

(2)   Left and Right Side Splits (In Case of Double Feed):

Since the only interfering currents are the zero sequence currents, the split between left and right sides is determined from the sequence network. Therefore, for case (a):

$$I_{L_a}/I_{Total\ Fault} = \frac{X_{TR_0} + (L-x)\ X_0}{X_{TL_0} + X_{TR_0} + LX_0} \quad , \quad (3)$$

$$I_{R_a}/I_{Total\ Fault} = \frac{X_{TL_0} + x\ X_0}{X_{TL_0} + X_{TR_0} + LX_0} \quad . \quad (4)$$

Case (b)

Assuming breaker B_R has opened, the only currents are those fed from G_L. In that case

$$Z_1 = X_{GL_1} + X_{TL_1} + x \cdot X_1 \ ,$$

$$Z_2 = X_{GL_2} + X_{TL_1} + x \cdot X_1 \ ,$$

$$Z_0 = X_{TL_0} + x \cdot X_0.$$

Therefore,

$$I_{Total\ Fault_b} = I_{L_b} = \sqrt{3} \times \frac{Base\ kVA}{Base\ kV} \left[\frac{100}{j\ [X_{GL_1}+X_{GL_2}+2X_{TL_1}+X_{TL_0}+x(2X_1+X_0)] + 3R} \right].$$

DOUBLE FEED — DOUBLE CIRCUIT. A typical double feed — double circuit arrangement is shown in the schematic diagram given in Figure 3-34.

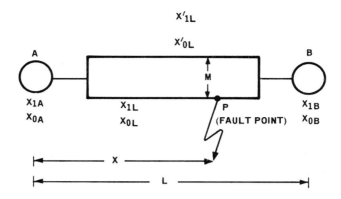

FIGURE 3-34

Double Feed — Double Circuit Schematic

In the following discussion, only case (a) will be considered.
Reactances X_{1L}, X_{0L}, X_{1L}', X_{0L}', and mutual inductance M are expressed
in percent per mile. Reactances X_{1A}, X_{0A}, X_{1B}, and X_{0B} are expressed
in percent.

Zero Sequence:

The zero sequence circuit may be analysed as shown in Figure
3-35.

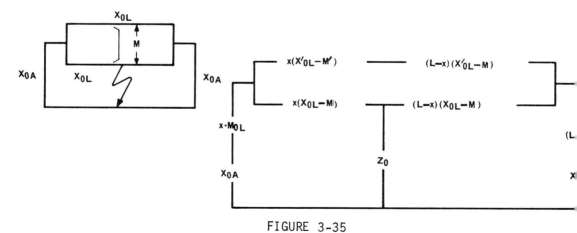

FIGURE 3-35

Zero Sequence Double Feed — Double Circuit

The zero sequence impedance Z_O is given by

$$Z_O = Z_{OB} + \frac{(Z_{OC} + X_{OB} + (L-x)M)\ (Z_{OA} + X_{OA} + x\,M)}{Z_{OA} + Z_{OC} + X_{OA} + X_{OB} + LM}\quad,$$

with $\quad Z_{OA} = \dfrac{x(X_{OL}-M)\ (X'_{OL}-M)}{X'_{OL} + X_{OL} - 2M}\quad,$

$$Z_{OB} = \frac{x(L-x)(X_{OL}-M)^2}{L\ (X_{OL}+X_{OL}-2M)}\quad,$$

$$Z_{OC} = \frac{(L-x)(X_{OL}-M)\ (X'_{OL}-M)}{X'_{OL} + X_{OL} - 2M}\quad.$$

Positive and Negative Sequences:

The positive and negative sequence circuit may be analysed as shown in Figure 3-36.

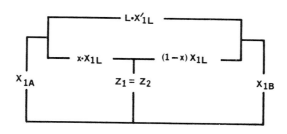

FIGURE 3-36

Positive and Negative Sequence Double Feed - Double Circuit

The impedances are given by

$$Z_1 = Z_2 = Z_{OB} + \frac{(Z_{OC} + X_{1B})\ (Z_{OA} + X_{1A})}{Z_{OA} + Z_{OC} + X_{1A} + X_{1B}}\quad,$$

with $\quad Z_{OA} = \dfrac{x \cdot X_{1L} \cdot X'_{1L}}{X_{1L} + X'_{1L}}\quad,$

$$Z_{\rho B} = \frac{(x\ X_{1L}) \cdot (L-x) \cdot (X_{1L})}{L\ (X_{1L} + X'_{1L})},$$

$$Z_{\rho C} = \frac{(L-x) \cdot (X_{1L} \cdot X'_{1L})}{X_{1L} + X'_{1L}}.$$

Residual Current Before Shielding:

This current is given by the following equation:

$$I\ (x) = 100\sqrt{3}\ \frac{Base\ kVA}{Base\ kV} \times \frac{1}{Zo + 2Z1} \times \frac{Z_{0C} + X_{0B} + (L-x)\ M}{Z_{0A} + Z_{0C} + X_{0A} + X_{0B} + LM}.$$

TWO CIRCUITS WITH ONE COMMON FEEDER POINT.   This case is shown in Figure 3-37.

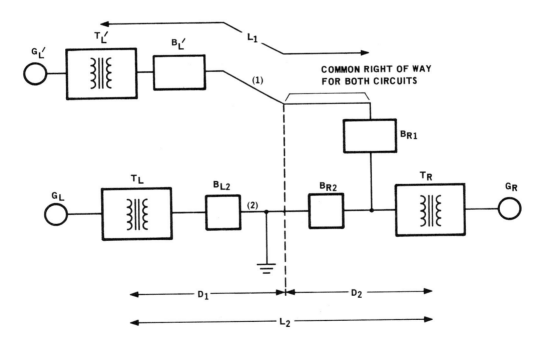

FIGURE 3-37
Circuits with One Common Feeder Point

The circuit shown in Figure 3-35 may be simplified for analysis as follows:

1) The contributions from G_L' may be ignored unless circuit (1) is much shorter than circuit (2), i.e., $L_1 < L_2/2$.

2) The return currents in the direction L-R may be approximated very closely as follows (provided G_L' may be ignored):

   a) Combine $G_L' + T_L' + Z_1$ with G_R and T_R (on same kV and kVA basis) to obtain an equivalent source.

   b) Distribute the mutual inductance over D_2 over the whole length of circuit (2), i.e., the zero sequence of circuit (2) becomes $X_0 - M_0 D_2/L_2$.

   c) Include $M_0 D_2$ in the zero sequence network of the circuit containing T_R.

   d) Make the total zero sequence of circuit (1) equal:

   $$L_1 \cdot X_{01} - D_2 M_0$$

   where X_{01} is the zero sequence line impedance of circuit (1).

   e) Return currents in the direction R-L are not affected.

TWO CIRCUITS ON SAME RIGHT OF WAY WITH NO ELECTRICAL CONNECTORS. This case is equivalent to the single circuit case with reduced zero sequence impedance. The reduction is partially offset by the shielding provided by the additional circuits.

3.14   L/G FAULT DATA

3.14.1   General

The line-to-ground (L/G) fault data was obtained by calculation using power systems parameters. The following assumptions were made:

a) Earth resistivity = 100 $\Omega \cdot$m
b) Generator reactances — Transient

Supplementary information was collected from published references.

Fault level distributions for the transmission network in Ontario were calculated by the Power Authority on the basis of typical lines by voltage class. Fault level distributions for the transmission network in Quebec were calculated from the various parameters of each line. In the Quebec case, only the reactance components were used in the calculations because the ratio $R_0/X_0 = R_1/X_1$, R_2/X_2, etc., was small (i.e., <1/3).

Fault levels of the distribution networks in the Richelieu and the Essex areas were obtained from power coordination maps in the two areas. The resistance components are not negligible in distribution networks and must be taken into account in the calculation of faults. However, overshoots are negligible and were ignored in the calculation.

3.14.2 Fault Resistances

Unless a fault occurs at or very close to a station, the resistance of the fault is not negligible when compared to the impedance of the circuit and can considerably decrease the levels of fault. The resistance in the path of a fault is essentially the arc resistance, R_A, in series with the grounding resistance, R_G, as seen from the point of fault. The combined resistance R_F appears in the zero sequence equivalent network as $3 R_F$.

The arc resistance is empirically given as[8]

$$R_A = \frac{44(kV)}{3I_0} \qquad \text{when kV} < 110$$

$$R_A = \frac{22(kV)}{3I_0} \qquad \text{when kV} \geq 110$$

where $3I_0$ is the fault current (L/G) calculated with zero fault resistance. That is, for an average L/G of $4 \times 10^3 A$, a 735 kV line would have an arc resistance of 4 Ω.

The grounding resistance depends on the tower footing resistance, the presence of skywires connected to the towers, the presence of counterpoises, and the location of the fault. For faults at a great distance from the source, the resistance of the grounding is much smaller than that of a single tower footing (see Figure 3-38).

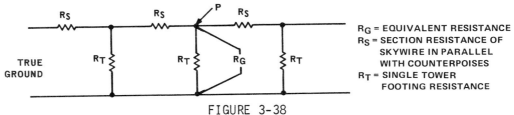

RG = EQUIVALENT RESISTANCE
RS = SECTION RESISTANCE OF
    SKYWIRE IN PARALLEL
    WITH COUNTERPOISES
RT = SINGLE TOWER
    FOOTING RESISTANCE

FIGURE 3-38

Skywire Tower Footing Model for Calculation
of System Grounding Resistance

Since there is usually a great spread in the values of R_T, acceptable grounding requires that R_G does not exceed 10 Ω when skywires are present[26].

Various organizations and authors have proposed average fault resistances that take into account both arc and grounding resistances [27,28,30,31,32]. After discussion with power company representatives, the limits quoted in Reference 13 were adopted for our calculations (see Table 3-11).

TABLE 3-11

Values of Fault Resistance
In the Case of L/G Fault[13]

FAULT LOCATION	FAULT RESISTANCE(R_F)
At Station	0 Ω
On Line With Skywires	15 Ω
On Line Without Skywires	50 Ω

These line fault values refer to metallic tower construction. In general, wood pole systems have larger grounding resistances. The breakdown on metallic and wooden pole systems is given in Table 3-12.

TABLE 3-12

Power System Configuration Breakdown

VOLTAGE	ONTARIO	QUEBEC
735 kV	–	Metallic
500 kV	Metallic	–
330 kV	–	Metallic
220 kV	Metallic	Metallic
110 kV	50% Metallic	No Data Available
Subtransmission and Distribution	Wood Poles	Wood Poles

3.14.3 Levels — Distribution with Respect To Distance

HYDRO QUEBEC TRANSMISSION. The results of the analysis of Hydro Quebec transmission are given in Figures 3-39 to 3-45.

In order to derive the statistical description shown on the graphs, it was assumed that the various parameters of the power circuits were not correlated. This assumption was confirmed by the analyses performed and corroborated through discussion with power representatives.

The future fault current distributions were made on the basis of maximum fault currents of breakers given by Hydro Quebec Systems Planning[29]. All the curves obtained closely followed the form

$$I(x) = \frac{V}{A + Bx} \quad \text{(single line, single phase equivalent)}$$

where

x = distance from source (miles)
V = phase to phase voltage (kV)
$I(x)$ = L/G fault current (kA)
A and B are constants dependent upon power system characteristics

The worst and typical A and B parameters for each voltage class are shown in Tables 3-13 and 3-14. The values given in these tables should be used in conjunction with the voltage class and not the actual operating voltage. The values of A parameters were derived from the maximum L/G fault current at the source to obtain the complete form. The fault resistance is not included in the values of A or B and is taken into account by modifying the equation as shown below:

$$I(x) = \frac{V}{j(A+Bx) + R_F}$$

where

R_F = fault resistance

TABLE 3-13

A Parameter Values — Hydro Quebec

VOLTAGE CLASS (kV)	PRESENT A (Ω)		FUTURE A (Ω)	
	TYPICAL	WORST	TYPICAL	WORST
735	88.5	88.5	64.0	64.0
330	25.0	18.55	17.10	10.05
220	20.0	9.65	13.4	6.8
110	8.05	4.03	5.45	3.11

TABLE 3-14

B Parameter Values — Hydro Quebec

Voltage Class	kV	110	220	330	735
Typical B	Ω/mile	3.39	3.00	2.55	3.14

ONTARIO HYDRO TRANSMISSION. The results of the analysis of Ontario
Hydro transmission are given in Figures 3-46 to 3-51. Because actual
line parameters were not available, the single phase single line model
was derived from the set of curves provided by the Ontario Hydro
Authority. Worst and typical A and B parameters are given in Tables
3-15 and 3-16.

TABLE 3-15

A Parameter Values — Ontario Hydro

VOLTAGE CLASS (kV)	TYPICAL A(Ω)	WORST A(Ω)
110	7.1	3.1
220	11.89	5.11
500	20.0	19.04

TABLE 3-16

B-parameter Values — Ontario Hydro

Voltage Class	kV	110	220	500
Typical B	Ω/mile	2.4	2.0	1.85

HYDRO QUEBEC DISTRIBUTION. The results of this analysis are given in
Figures 3-52 and 3-53. As in the case of transmission, no overshoot
and zero fault resistance were assumed.

However, as resistance is not negligible in the case of
distribution networks, the line impedance rather than reactance
was used. The curves follow the relationship

$$I(x) = \frac{V}{j(A+X\cdot x) + R\cdot x}$$

where

$I(x)$ = fault current (kA)
A = source impedance (ohms)
x = distance from source (miles)
X = line reactance (ohms/mile)
R = line resistance (ohms/mile)
V = phase to phase voltage (kV)

If the fault resistance has a value of R_F instead of zero, the above
equation can be modified as follows:

$$I(x) = \frac{V}{j(A+X\cdot x) + R\cdot x + R_F}$$

148

The worst and typical values of parameter A, R, and X for the 10 kV distribution class, which is used for most of the Hydro Quebec distribution network studied, is shown in Table 3-17.

TABLE 3-17

10 kV Class Parameters — Hydro Quebec Distribution System

PARAMETER	UNITS	TYPICAL VALUE	WORST VALUE
A	Ω	3.11	1.96
R	Ω/mile	1.69	1.69
X	Ω/mile	1.82	1.82
Fault Current Level	amps	660	3300

ONTARIO HYDRO DISTRIBUTION. The results of this analysis are given in Figures 3-54 and 3-55.

The worst and typical values of parameters A, R and X for the 10 kV distribution class, widely used by Ontario Hydro distribution system, are given in Table 3-18.

TABLE 3-18

10 kV Class Parameters, Ontario Hydro Distribution System

PARAMETER	UNITS	TYPICAL VALUE	WORST VALUE
A	Ω	3.53	2.36
R	Ω/mile	1.325	1.325
X	Ω/mile	1.37	1.37
Fault Current level	amps	577	1153

149

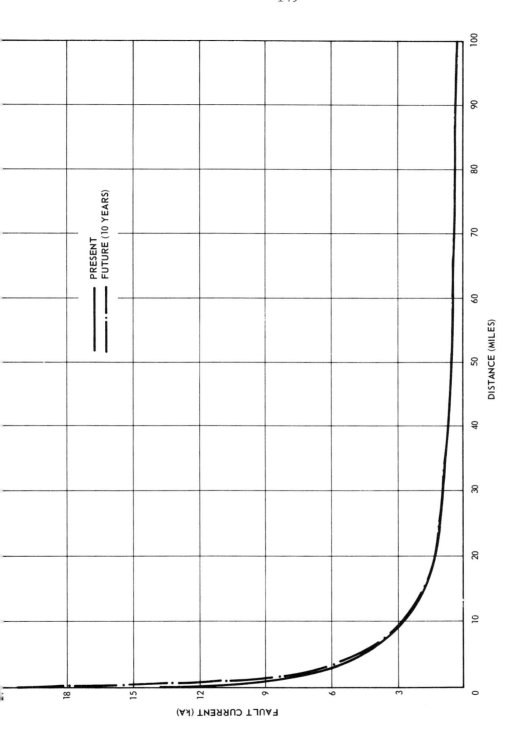

FIGURE 3-39

Single L/G Fault Current Versus Distance from the Source —
Typical Case for 110 kV Hydro Quebec Lines

150

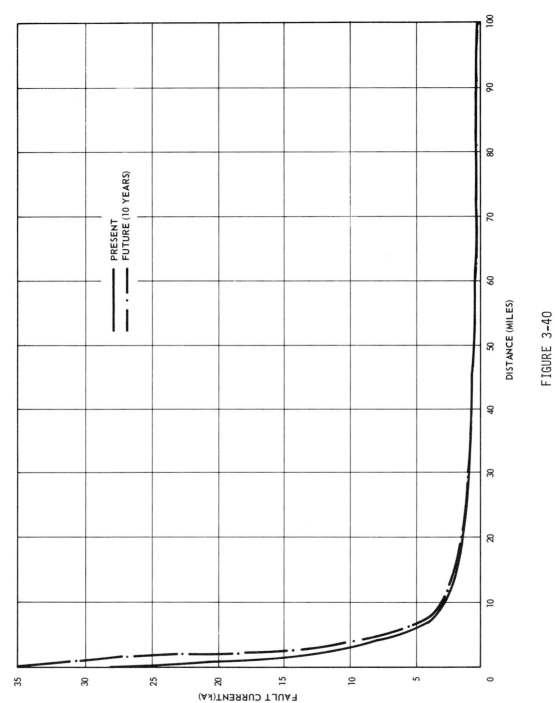

FIGURE 3-40

Single L/G Fault Current Versus Distance From the Source —
Worst Case for 110 kV Hydro Quebec Lines

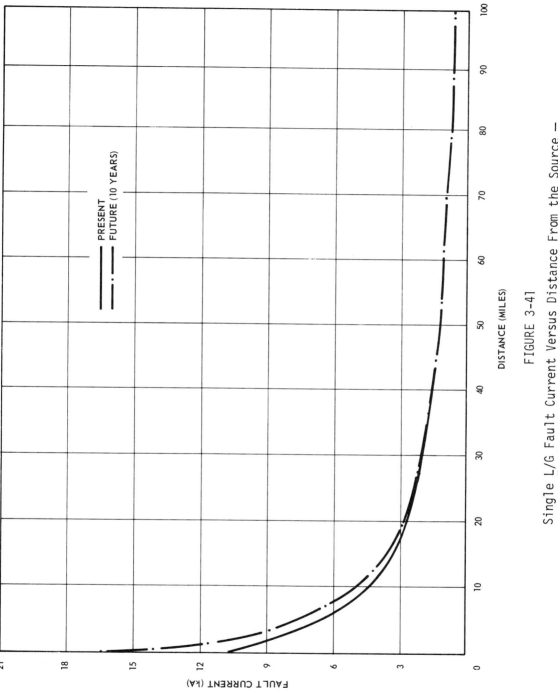

FIGURE 3-41

Single L/G Fault Current Versus Distance From the Source —
Typical Case For 220 kV Hydro Quebec Lines

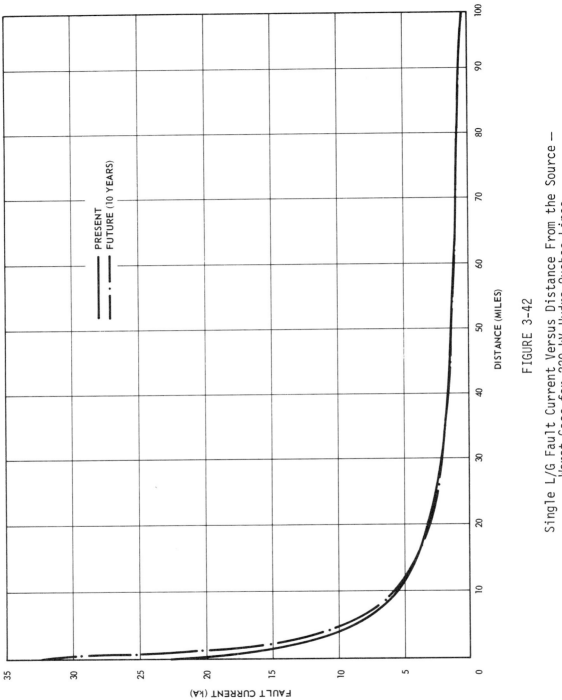

FIGURE 3-42

Single L/G Fault Current Versus Distance From the Source —
Worst Case for 220 kV Hydro Quebec Lines

153

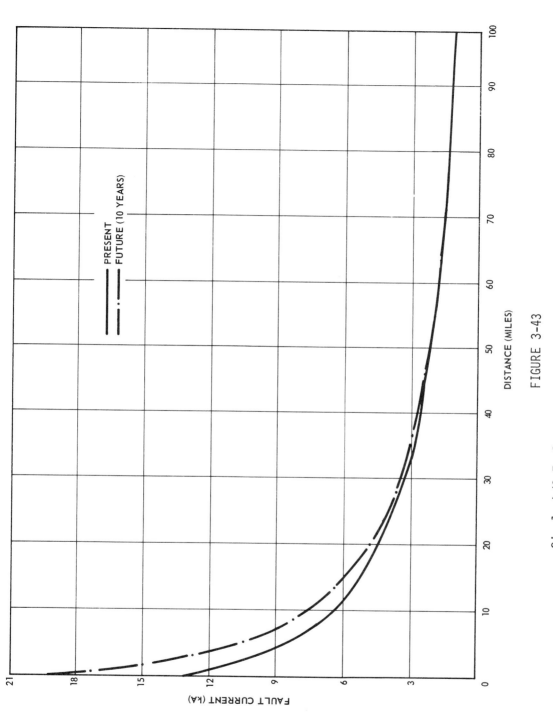

FIGURE 3-43

Single L/G Fault Current Versus Distance from the Source —
Typical Case For 330 kV Hydro Quebec Lines

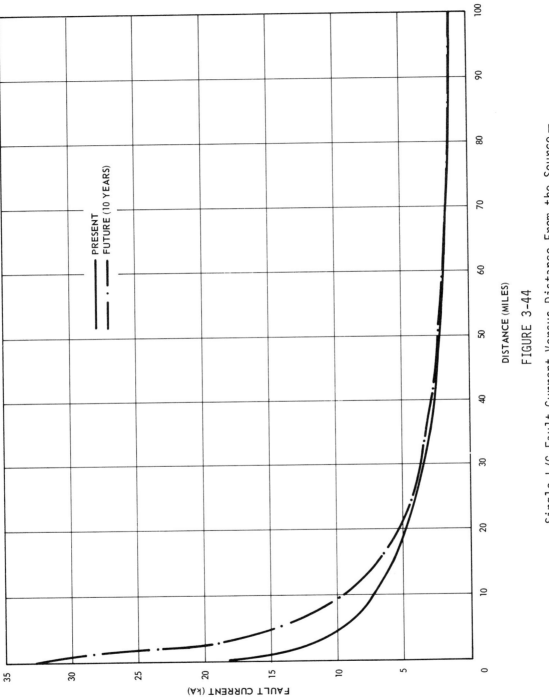

DISTANCE (MILES)

FIGURE 3-44

Single L/G Fault Current Versus Distance From the Source —
Worst Case for 330 kV Hydro Quebec Lines

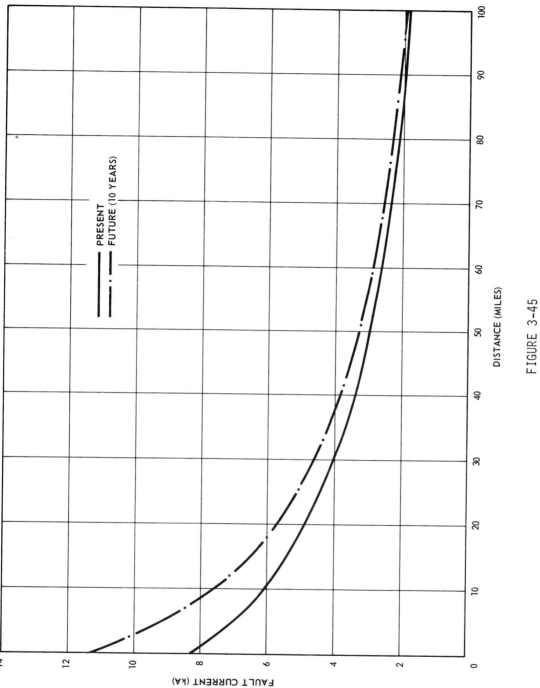

FIGURE 3-45

Single L/G Fault Current Versus Distance From the Source —
Typical Case for 735 kV Hydro Quebec Lines

156

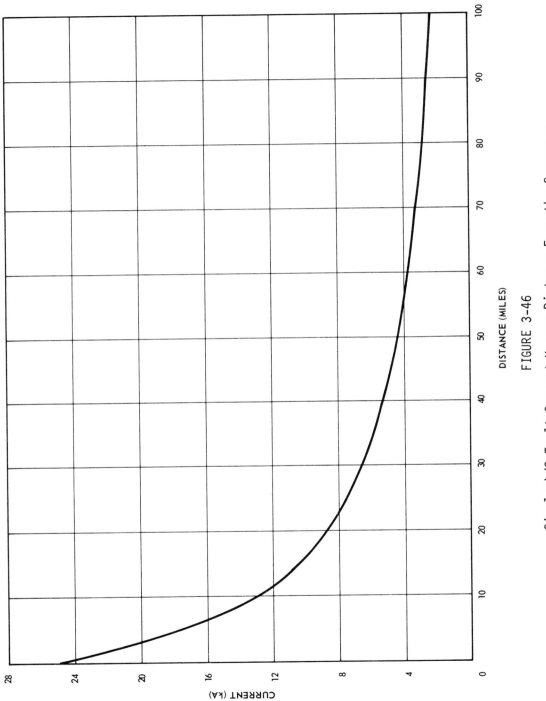

DISTANCE (MILES)

FIGURE 3-46

Single L/G Fault Current Versus Distance From the Source —
Typical Case for Present 500 kV Ontario Hydro Lines

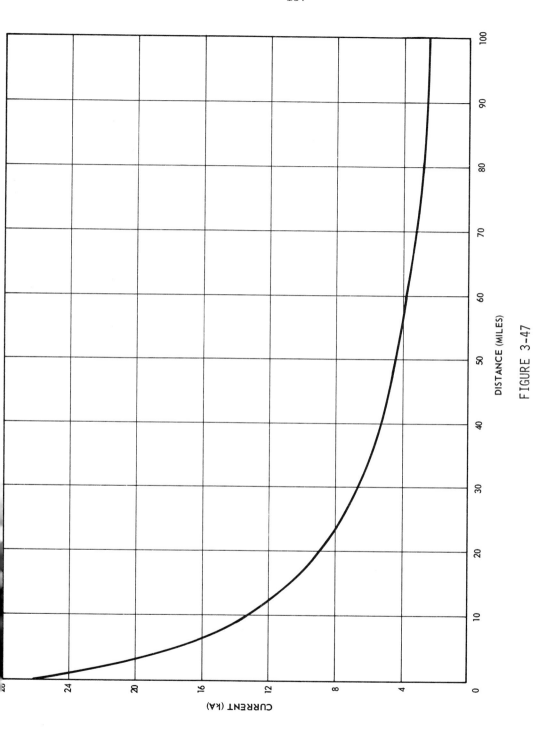

DISTANCE (MILES)

FIGURE 3-47

Single L/G Fault Current Versus Distance From the Source —
Worst Case for Present 500 kV Ontario Hydro Lines

158

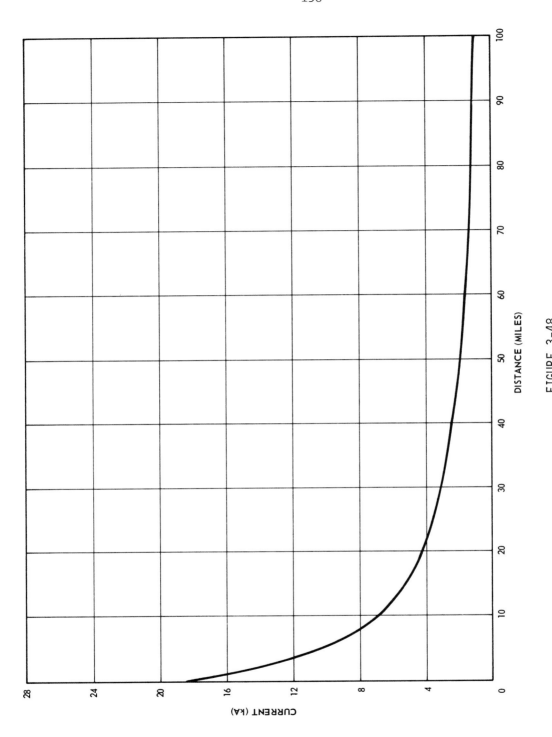

FIGURE 3-48

Single L/G Fault Current Versus Distance From the Source —
Typical Case for Present 220 kV Ontario Hydro Lines

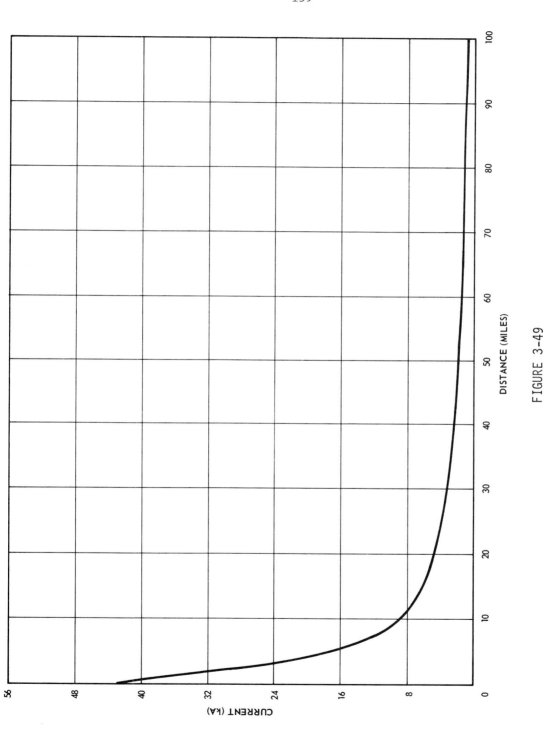

FIGURE 3-49

Single L/G Fault Current Versus Distance From the Source —
Worst Case for Present 220 kV Ontario Hydro Lines

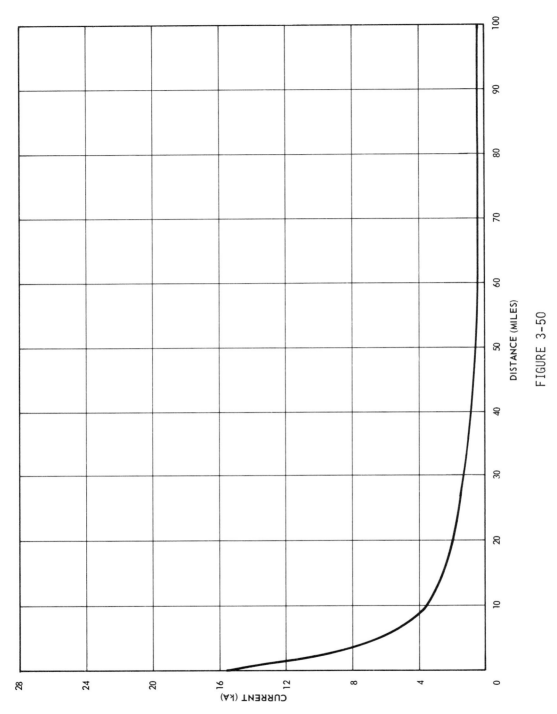

DISTANCE (MILES)

FIGURE 3-50

Single L/G Fault Current Versus Distance From the Source —
Typical Case For Present 110 kV Ontario Hydro Lines

161

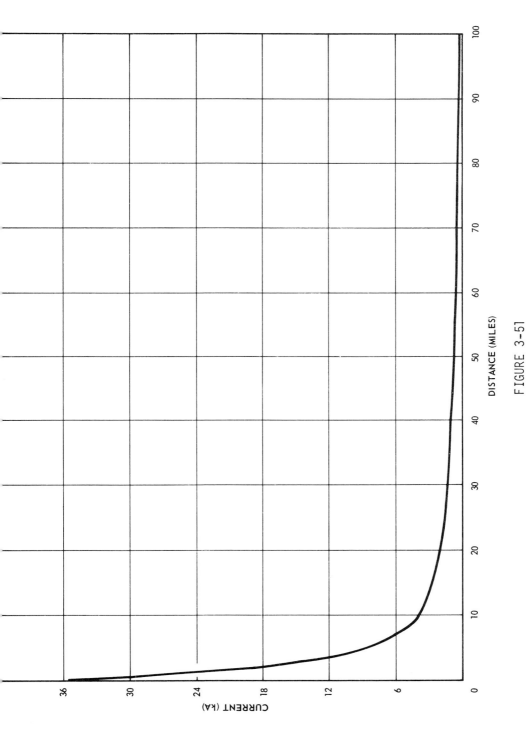

DISTANCE (MILES)

CURRENT (kA)

FIGURE 3-51

Single L/G Fault Current Versus Distance From the Source —
Worst Case for Present 110 kV Ontario Hydro Lines

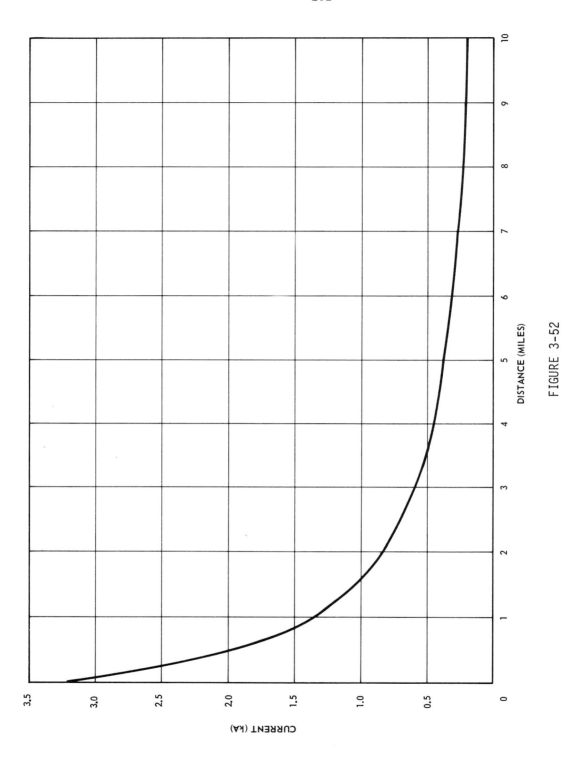

FIGURE 3-52

Single L/G Fault Current Versus Distance From the Source —
Typical Case for Present 10 kV Hydro Quebec Lines

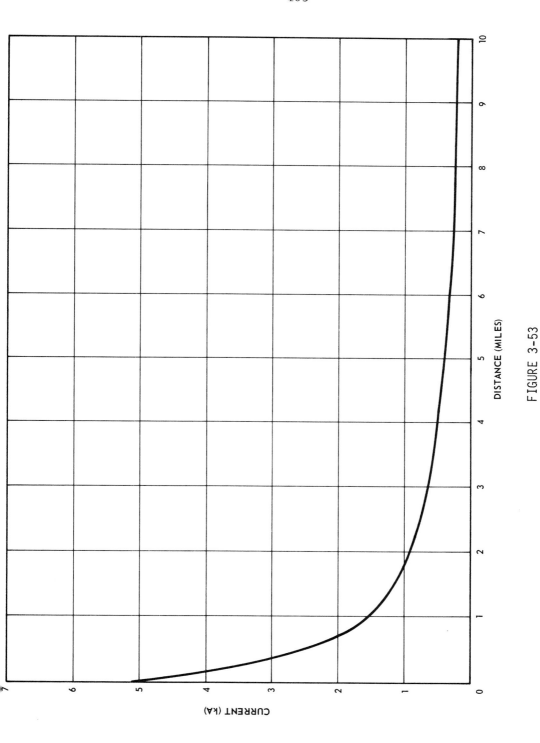

FIGURE 3-53

Single L/G Fault Current Versus Distance From the Source —
Worst Case for Present 10 kV Hydro Quebec Lines

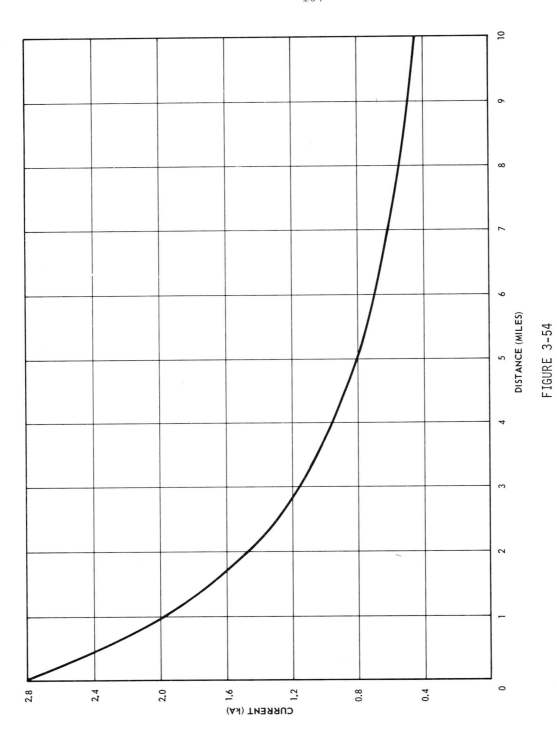

FIGURE 3-54

Single L/G Fault Current Versus Distance From the Source —
Typical Case For Present 10 kV Ontario Hydro Lines

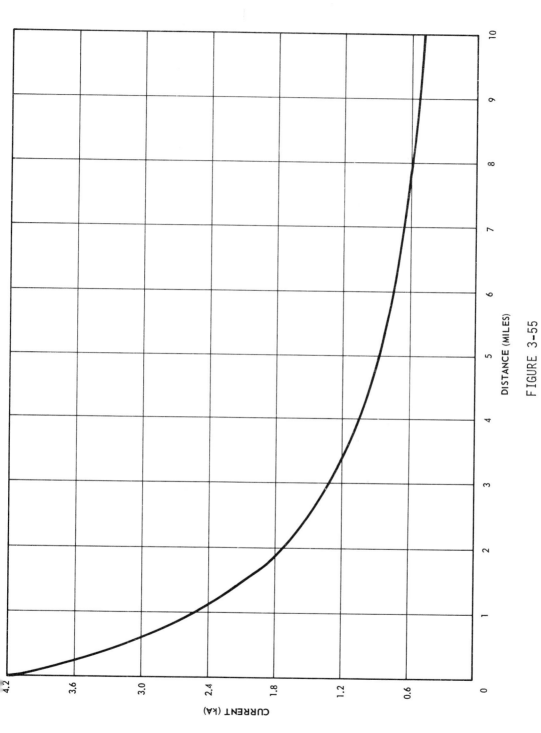

FIGURE 3-55

Single L/G Fault Current Versus Distance from the Source—
Worst Case for Present 10 kV Ontario Hydro Lines

3.14.4 Power System Protection — Fault Clearing Timing Sequence

HYDRO QUEBEC TRANSMISSION. The typical timing sequence of power system protective devices shown in Figure 3-56 applies to the Hydro Quebec transmission network.

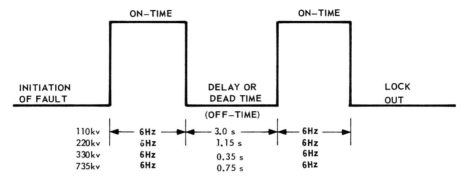

FIGURE 3-56

Typical Timing Sequence of Power System
Protective Devices — Hydro Quebec

In other words, induced voltage due to fault current will last typically for one-tenth of a second and if the fault is not cleared in the first try, a voltage of equal magnitude will be induced for another one-tenth of a second after a period of time defined as the 'dead time' above. After the second trial the devices lockout if the fault remains uncleared.

Table 3-19 summarizes the possible minimum and maximum on and off times for each voltage class for the Hydro Quebec Transmission network.

TABLE 3-19

Timing Sequence — Hydro Quebec Transmission Network

VOLTAGE CLASS kV	ON-TIME IN SECONDS		OFF-TIME IN SECONDS	
	MIN	MAX	MIN	MAX
735	0.1	0.33	0.75	0.75
330	0.1	0.5	0.2	1.5
220	0.1	1.0	0.17	10.0
110	0.1	1.0	0.5	5.0

ONTARIO HYDRO TRANSMISSION. The typical on-off cycle for Ontario Hydro transmission system is shown in Figure 3-57.

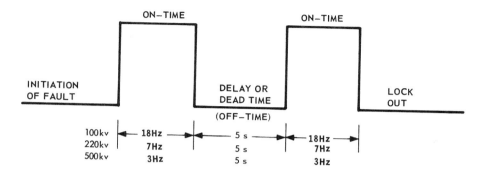

FIGURE 3-57

Typical Timing Sequence of Power System
Protective Devices - Ontario Hydro

Here again, the induced voltage due to fault may last three-tenths of a second for each of the two times with a time interval of 5 seconds. No worst case data was available.

HYDRO QUEBEC DISTRIBUTION. Figure 3-58 shows the results of statistical analysis on the 10 kV class on the Hydro Quebec distribution system.

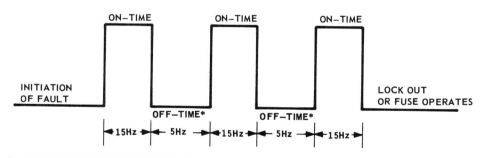

* THE OFF TIMES VARY CONSIDERABLY FOR ELECTROMECHANICAL
AND ELECTRONICALLY PROGRAMMED RECLOSURES.

FIGURE 3-58

Typical Timing Sequence of Hydro Quebec Distribution System

Possible maximum and minimum on and off times (total for sequence)
are given below:

            Number of cases analysed   = 122
            Average number of reclosures = 3-4
            Total minimum on-time   = 0.15 s
            Total minimum off-time   = 0.15 s
            Total maximum on-time   = 8.0 s
            Total maximum off-time   = 1.0 s

ONTARIO HYDRO DISTRIBUTION. The statistical analysis yielded the
results shown in Figure 3-59 for the 10 kV class on the Ontario
Hydro distribution system.

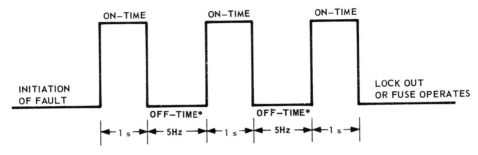

*** THE OFF TIMES VARY CONSIDERABLY FOR ELECTROMECHANICAL
AND ELECTRONICALLY PROGRAMMED RECLOSURES.**

FIGURE 3-59

Typical Timing Sequence for Ontario Hydro Distribution System

Possible maximum and minimum on and off times (total for sequence)
are given below:

            Number of cases analysed   = 71
            Average number of reclosures = 3-4
            Total minimum on-time   = 0.8 s
            Total minimum off-time   = 0.16 s
            Total maximum on-time   = 6.5 s
            Total maximum off-time   = 1.05 s

3.14.5 Power Systems Transmission and Distribution

A. Statistics on the Current-Time Relationship

HYDRO QUEBEC DISTRIBUTION. Linear regressions performed to correlate fault current levels and corresponding on-off times for the 10 kV class in the Hydro Quebec distribution network gave the following results:

>   No. of samples analyzed = 122
>
>   Average total time (sum of on and off times) = 1.20 s
>
>   Best Linear Equation: $t = 1.60 - 0.00051\ I$
>> where
>> t = total time (s)
>> I = L/G fault current (A)
>
>   Correlation coefficient = 0.188

ONTARIO HYDRO DISTRIBUTION. The corresponding analysis for the Ontario Hydro 10 kV distribution class gave the following results:

>   No. of samples analyzed = 70
>
>   Average total time (sum of on and off times) = 4.0 s
>
>   Best Linear Equation: $t = 9.1 - 0.0084\ I$
>> where
>> t = total time (s)
>> I = L/G fault current (A)
>
>   Correlation coefficient = 0.609

B. Field Strength and Induced Voltages

Whenever there is an exposure of a communication network to a power network, a field* affecting the communication networks may be defined as the product of the power system current and the coupling impedance per unit length between the two networks. During abnormal conditions, the power system current may be the fault current for a line to ground fault.

The induced voltage is the product of the power system current and total coupling impedance for a particular exposure.

The results of statistical analysis for different systems are presented in Tables 3-20 and 3-21.

---

* Also known as disturbing field or primary field.

170

TABLE 3-20

Hydro Quebec Transmission and Distribution (T&D) Network

VOLTAGE CLASS (kV)	INDUCED VOLTAGE(V)		FIELD (V/kft)	
	TYPICAL	WORST	TYPICAL	WORST
735	175	1938	26.6	92.5
330	80	3809	3.7	105.3
220	435	5056	24.6	438.1
110	150	2160	20.8	330.5
10 (DIST)*	209	1427	23.4	134.7

* Does not include joint use situations.

TABLE 3-21

Ontario Hydro Transmission and Distribution (T&D) Network

VOLTAGE CLASS (kV)	INDUCED VOLTAGE(V)		FIELD(V/kft)	
	TYPICAL	WORST	TYPICAL	WORST
500	153	1890	19	310
220	353	5287	22	469
110	126	1698	14	233
10 (DIST)*	138	731	10.2	63

* Does not include joint use situations.

3.15   GEOGRAPHICAL PARAMETERS

3.15.1 General

Geographical parameters are defined as those non-electrical parameters which are associated with a communication network due to its physical presence in a particular environment. Examples are separation from power networks, exposure length, exposure density, etc.

Tables 3-22 and 3-23 show the results of the statistical analysis on Hydro Quebec and Ontario Hydro transmission and distribution systems. The following parameters were considered:

1. Ground Resistivity
2. Exposure Length
3. Separation

TABLE 3-22

Geographical Parameters — Hydro Quebec T&D Network

VOLTAGE CLASS (kV)	GROUND RESISTIVITY RANGE (Ω-m)	EXPOSURE LENGTH (MILES)		EQUIVALENT SEPARATION (ft)	
		TYPICAL	WORST	TYPICAL	WORST
735	100-500	4.17	30.5	5021	3391
330	100-500	4.3	67.0	11900	1690
220	100-500	8.0	67.0	5888	719
110	100-500	2.77	28.9	5120	241
10 (DIST)*	100-500	1.33	7.95	1022	200

* Does not include joint use situations.

TABLE 3-23

Geographical Parameters — Ontario Hydro T&D Network

VOLTAGE CLASS (kV)	GROUND RESISTIVITY RANGE (Ω-m)	EXPOSURE LENGTH (MILES)		EQUIVALENT SEPARATION (ft)	
		TYPICAL	WORST	TYPICAL	WORST
500	50-250	1.22	3.66	6200	4255
220	50-250	4.09	43.5	6000	620
110	50-250	5.27	24.6	7206	593
10 (DIST)*	50-250	1.0	6.5	1058	200

* Does not include joint use situations.

3.15.2 Power Exposure Density

The power exposure density for a particular voltage class is defined as the sum of exposure lengths to power lines belonging to that voltage class divided by the route length of exposed communication cable or line. The units used for power exposure density here will be miles of power line per mile of communication cable.

Tables 3-24 and 3-25 show the results of the statistical analysis on Hydro Quebec and Ontario Hydro transmission and distribution systems. The figures represent ratio of power and telephone network population in selected Quebec and Ontario areas.

TABLE 3-24

Hydro Quebec T&D Network
Power Exposure Density (Powerline length/mile of
Communication Cable Route)

VOLTAGE CLASS (kV)	TYPICAL	WORST
735	1.25	2.20
330	0.95	2.40
220	0.62	1.56
110	0.62	3.10
10 (DIST)*	1.666	2.79

* Does not include joint use situations.

TABLE 3-25

Ontario Hydro T&D Network
Power Exposure Density (Powerline length/mile of
Communication Cable Route)

VOLTAGE CLASS (kV)	TYPICAL	WORST
500	0.12	0.6
220	0.18	1.32
110	0.285	0.83
10 (DIST)*	1.26	2.09

* Does not include joint use situations.

3.16   OTHER CONSIDERATIONS

3.16.1 Power System Reliability

Previous historical and design data from various sources was manipulated to give the curves of Figure 3-60 which show typical system outage rated (due to lightning and other faults) per 100 miles of power line length per year versus operating voltage.

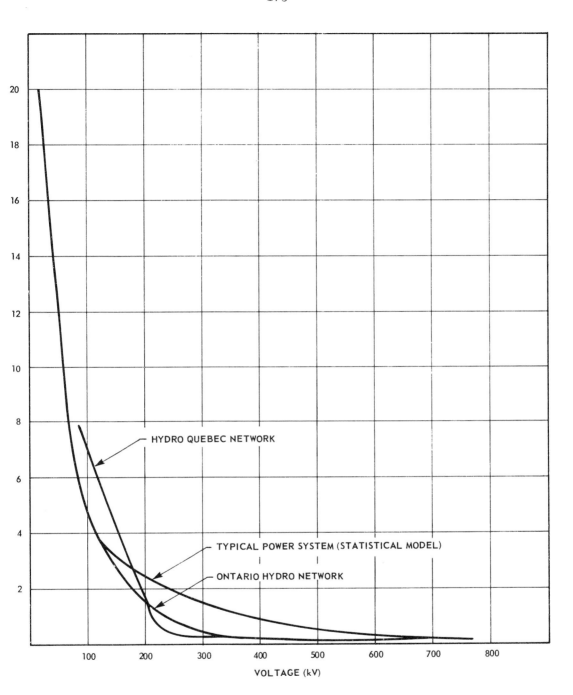

FIGURE 3-60

Total Outages Rate 'R' of Power Lines in Yearly Occurrence
Per 100 miles versus Voltage Classes in kV

It is obvious from these curves that the higher the voltage of operation, the more reliable the system becomes, which indeed should be the case.

3.16.2 Skywire Shielding

Power lines are sometimes equipped with a ground wire and from the communication engineer's point of view, it is important that the distribution of current in skywires during fault conditions be known, so that the effective shielding factor due to the presence of skywire can be calculated.

Desieno, Marchenko and Vassell[33] published their work in 1970 which describes a method for accurate determination of the transmission line current through ground wires during fault conditions.

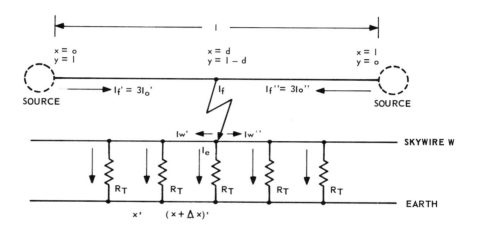

FIGURE 3-61
Transmission Line Current Through Ground Wires

Consider one phase of a transmission line with ground wire as shown in Figure 3-61. Let the towers be spaced at a distance Δx apart and let the footing resistance of each tower be R_t. The difference between the ground wire fault current $I_w(x + \Delta x)'$ at midspan $(x + \Delta x)$ and the ground wire fault current $I_w(x)'$ is the current I_e' which flows through the tower footing resistance R_t. Consider only the left hand side from the point of fault:

$$I_w(x + \Delta x)' - I_w(x)' = I_e' \qquad (1)$$

Also the current into earth I_e' can be expressed as the voltage of the ground wire V_w' multiplied by the tower footing conductance $G_t = 1/R_t$:

$$I_e' = G_t V_w' = \frac{1}{R_T} V_w'. \tag{2}$$

By combining (1) and (2) and dividing by Δx, the following relation is obtained:

$$\frac{I_w(x+\Delta x)' - I_w(x)'}{\Delta x} = \frac{G_t}{\Delta x} V_w'. \tag{3}$$

But $G_t/\Delta x$ = tower footing conductance/mile = G (say);

then

$$G = (G_t/\Delta x) \tag{4}$$

and

$$\frac{I_w(x+\Delta x)' - I_w(x)'}{\Delta x} = G V_w' . \tag{5}$$

To obtain a differential equation from (5), it is necessary to take the limit as Δx approaches zero, but in this case it is tower spacing and cannot reach zero. However Δx is usually very small compared to the line length ℓ and hence the limit may be defined.

Thus,

$$\frac{dI_w'}{dx} = G V_w' . \tag{6}$$

The voltage drop across the ground wire betweeen points $(x+\Delta x)$ and Δx is

$$V_w(x+\Delta x)' - V_w(x)' = I_w' Zww \Delta x - I_f' Zaw \Delta x \tag{7}$$

where

Zww = self impedance of ground wire per unit length
Zaw = mutual impedance between faulted phase and ground wire per unit length

Rearranging (7) gives

$$\frac{V_w(x+\Delta x)' - V_w(x)'}{\Delta x} = (I_w' Zww - I_f' Zaw). \tag{8}$$

Taking the limit,

$$\frac{dV_w'}{dx} = I_w'Zww - I_f'Zaw. \tag{9}$$

Differentiating (6),

$$\frac{d^2I_w'}{dx^2} = G\frac{dV_w'}{dx} = GZwwI_w' - GZawI_f'. \tag{10}$$

Solution of (10) gives

$$I_w'(x) = A_1'\varepsilon^{rx} + A_2'\varepsilon^{-rx} + \frac{Zaw}{Zww} I_f' \tag{11}$$

and

$$V_w'(x) = \frac{1}{G} A_1'r\varepsilon^{rx} - \frac{1}{G} A_2'r\varepsilon^{-rx}, \tag{12}$$

where

$$r = GZww \text{ and } A_1', A_2' \text{ are arbitrary constants.}$$

Assuming that at the system terminals, there is good grounding, then

$$V_w'(0) = \frac{1}{G} A_1'r - \frac{1}{G} A_2'r = 0,$$

$$\text{giving } A_1' = A_2' = A' \text{ (say).}$$

Then (11) and (12) become

$$I_w'(x) = 2A' \cos hrx + \frac{Zaw}{Zww} I_f' \tag{13}$$

and

$$V_w'(x) = \frac{2rA'}{G} \sin hrx . \tag{14}$$

Similar equations can be worked out from the right-hand end, yielding:

$$I_w''(y) = 2A'' \cos hry + \frac{Zaw}{Zww} I_f'' \tag{15}$$

and

$$V_w''(y) = \frac{2rA''}{G} \sin hry, \tag{16}$$

at $x = d$, $y = 1-d$.

Then,

$$V_w'(d) = \frac{2rA'}{G} \sin hrd \qquad (17)$$

and

$$V_w''(1-d) = \frac{2rA''}{G} \sin hr(1-d). \qquad (18)$$

But, since

$$V_w'(d) = V_w''(1-d) = V_w(say),$$

therefore,

$$A'' = \frac{A' \sin rd}{\sin hr(1-d)} \cdot \qquad (19)$$

Also at x = d,

$$I_e = \frac{V_w}{R_t}, \qquad (20)$$

and $I_f = I_w' + I_w'' + I_e$, giving

$$I_f = 2A' \cos hrd + 2A'' \cos hr(1-d) + \frac{Zaw}{Zww}(I_f' + I_f'') + \frac{2rA'}{GR_t} \sin hrd.$$

Using (19) for the relationship between A' and A'', we get a final result:

$$I_w' = \frac{I_f(1-\frac{Zaw}{Zww}) \cos hrx}{\cos hrd \left[1+(\tan hrd)\frac{r}{GR_t}+\frac{1}{\tan hr(1-d)}\right]} + \frac{Zaw}{Zww} I_f' \qquad (21)$$

and

$$I_w'' = \frac{I_f(1-\frac{Zaw}{Zww} \tan hrd \cos hry}{[\sin hr(1-d)] [1+(\tan hrd)\frac{r}{GR_t} + \frac{1}{\tan hr(1-d)}]} + \frac{Zaw}{Zww} I_f''. \qquad (22)$$

The above equations can be applied to a radial line with fault contribution from one side only by letting $I_f' = I_f$ and $I_f'' = 0$.

3.16.3 Problems in Metropolitan Areas[34]

When a fault occurs on the power line between a phase and earth, the current unbalance may reach many thousands of amperes in magnitude depending on the power circuit parameters. The potential induced on the telephone wire may then reach thousands of volts, depending mainly on the length of the exposure and the separation between the two routes.

In the simple case of a power line fault at or beyond the ends of a uniform exposure with protection provided at the ends of the exposure, the highest potentials on the telephone lines will occur at its ends. In practice, exposures are seldom uniform and the power faults may occur at points within the exposure. In such cases, the maximum line potential may occur at some point other than its ends, and this situation should be considered in protection design.

Telephone cables are liable to the same type of induction as open wires. If the cable is provided with a metallic sheath, the sheath also is subjected to the same induction. This current will, in turn, induce an EMF in the conductors of the cable, which is in phase opposition to the directly induced potential. This is termed shielding and its effectiveness is improved by:

1) Increasing the mutual inductance between sheath and conductors.

2) By reducing the resistance of the sheath.

The induced voltage is also a function of the resistivity of the earth in the vicinity of the exposure. Since the earth is a very poor conductor, the current returning through the ground from a power fault to the neutral point of the supply transformer must therefore spread to great depths to achieve a low resistance path by increasing the cross-sectional area.

In metropolitan areas, where numerous metallic pipes and other conductor are laid close to the earth's surface, a portion of the earth return current will flow in these conductors reducing the effective resistivity of earth. Those conductors close to either power or telephone routes will also exert a shielding effect due to induced current. The net result is a considerable reduction in the magnitude of the induced voltage on telephone conductors in metro areas — normally to much less than that calculated for the same parameters in rural areas. Despite the mitigating effects of buried conductors, and the short exposures which are normal, induced potentials can still be high in metropolitan areas due to the high fault currents which are likely and care should be taken to decide if the elimination of some protective devices is warranted for new built-up areas.

The power fault current is dependent on the source impedance from the generators to the supplying sub-stations, the power capacity of the transformers supplying the line, and the impedance of the line to the point of the fault. In urban areas all factors are favorable for high fault currents.

Power lines in metropolitan areas are normally protected by both over-current and earth leakage circuit breakers. The characteristics of relays are such that the time of operation is inversely proportional to the magnitude of the fault. Thus, high induced voltages are short in duration.

A large percentage of power faults occur between phases due to birds and other flying objects — and these have negligible effect on telephone circuits as the fault currents balance in the phase wires. Only faults to earth cause substantial external induction, and the majority of these are due to lightning. At such times, the maintenance staff is unlikely to be working in the field; consequently the shock hazard is not significant. Furthermore, a large portion of line work is performed in the area close to central office as this is the area of maximum cable density and usually of maximum subscribers' density. Since telephone lines are earthed at the exchange by MDF protection potential buildup tends to occur towards the extremity of the cable remote from the exchange. Considering the factors mentioned above, the probability of dangerous exposure for line staff is very low in metropolitan areas.

Conversely telephone plant and associated equipment are relatively more exposed to dangerous potentials in the event of a power line fault.

Protection for these facilities should be considered on the basis of system economic factors.

3.17   DC OFFSET — OVERSHOOT

3.17.1 General

When a fault occurs on a power circuit, the fault current goes from an initial high value, through a transient state, to the permanent short-circuit value (which is the lowest). This phenomenon is due to two main effects:

a)   transient generation due to the sudden connection of a voltage source to an R/L circuit,

b)   variation of the positive reactances of the rotating components of the power network.

Because, from a communications engineer's point of view, L/G faults are the most severe and the most frequent, only L/G variations will be described (Figure 3-62).

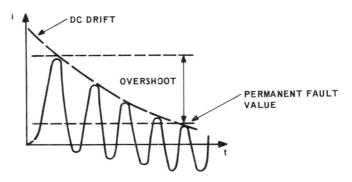

FIGURE 3-62

Variation of L/G Fault Current With Time

3.17.2 Significance of Parameters

Let the fault impedance Z be given by

$$Z = Z_0 + Z_1 + Z_2 + 3R_F,$$

where

Z_0 and Z_2 are the zero and negative sequence impedances, respectively, seen from the fault point,

R_F is the resistance of the fault, and

$Z_1(t)$ is the positive sequence impedance seen from the fault point.

$Z_1(t)$ varies with time, and goes from a low value (subtransient reactance) to a high value (steady state reactance). The times that are of interest in determining peaks are short enough to consider $Z_1(t) = Z_1$ = etc. In a subsequent paragraph, a comparison will be made between subtransient and steady-state conditions.

Let

$e = E\sqrt{2} \sin(\omega t + \psi)$, the emf applied at the inception of the fault,

ψ = the angle at the time of closure,

$\rho = \arctan \omega L/R$, where $R = Re(Z)$, and $\omega L = Im(Z)$,

η = the shielding factor provided by the skywires and/or counterpoises of the power circuit.

The sudden interconnection of a voltage, e, to an R/L circuit gives rise to a current, i, that is given by

$$i = \frac{E\sqrt{2}}{Z} \{\sin(\omega t + \psi - \rho) - \sin(\psi - \rho) \ e^{-(R/L)t}\}$$

$$= \frac{k}{Z} \{\sin(\omega t + \psi - \rho) - \sin(\psi - \rho) \ e^{-(R/L)t}\} \ . \tag{1}$$

In the case of magnetic coupling, the induced voltage on the communication circuit is proportional to the derivative of the current:

$$E_{ind} = \frac{K}{Z} \{\omega\cos(\omega t + \psi - \rho) + \frac{R}{L} \sin(\psi - \rho) \ e^{-(R/L)t}\}. \tag{2}$$

Equation (1) determines the influence of resistive coupling, and Equation (2) determines the influence of magnetic coupling.

Since Z is a function of time in this case, the severities of the overshoot and dc drift are determined by the ratio $\omega L/R = X/R$ and the angle ψ.

3.17.3 Transmission Circuits (≥100 kV)

Because of the speed of de-energization, only the first few cycles are of interest when considering transmission circuits. The reactive component of these circuits (including load, line, transformers, and generators) is much greater than the resistive component, particularly near the source.

Data collected for the major power networks in Bell Canada territory gave a breakdown of the ratio of X/R at the terminals as shown in Tables 3-26, 3-27, and 3-28. See also Figure 3-63.

TABLE 3-26

Typical X/R Values

VOLTAGE CLASS (kV)	X/R RANGE	TYPICAL X/R AT TERMINALS
735	18.9 − 20.4	19.7
500	13.6 − 16.5	16.5
330	6−40	18
220	2−25	9
110	3−26	14

Maxima are in the order of 40, and typical values range from 10 to 20. On the line, the X/R value decreases with separation from the source and, at about 20 miles, is approximately equal to the X/R of the line itself, with a maximum value of 7.

The presence of a resistance in the fault path substantially decreases the X/R ratio, particularly near the source (see Table 3-30).

For faults on the line, the angle ψ is usually close to $\pi/2$. This is because the faults are usually caused by a breakdown of the weakest dielectric path, and this happens when the voltage is maximum, i.e., when $\psi = \pi/2$. At stations, faults can be sometimes caused by human error, and thus the angle ψ can be random. A maximum of twice the steady state short circuit value could be reached in this case, but a design figure of 1.8 is usually adequate.

TABLE 3-27

X/R Ratios for Hydro Quebec Transmission Lines

VOLTAGE CLASS	735 kV	330 kV	220 kV	110 kV
Terminal, Range: X/R	18.9 – 20.4	6.1 – 39.8	6.7 – 25	2.7 – 25.9
Terminal, Typical: X/R	19.7	18.0	10.0	7.9
X – R (%)	2.017 – 0.108	2.93 – 0.16	6.5 – 0.65	17.8 – 2.2
Terminal, Worst: X/R	sample size	39.8	25.0	25.9
X – R (%)	too small	2.79 – 0.07	3.5 – 0.14	7.26 – 0.28
Line, Range: X/R	NA	5.0 6.8	4.0 5.2	4.0 5.5
Line, Typical: X/R	6.2	6.6	4.8	4.1
X – R (% mile)	0.52 – 0.084	0.4 – 0.06	0.87 – 0.18	2.6 – 0.64
Line, Worst: X/R	NA	6.8	5.2	5.5
X – R (% mile)	NA	0.38 – 0.056	0.88 – 0.17	2.2 – 0.4
15Ω Fault (%) (See Note 1)	0.28	1.58	3.1	10.4

NOTES:

1. A value of three times that indicated should be used to account for the effect of the fault resistance.

2. X/R at terminals specified in the subtransient condition.

TABLE 3-28

X/R Ratios for Ontario Hydro Transmission and Subtransmission Lines

VOLTAGE CLASS	500 kV	220 kV	110 kV	44 kV
Terminal, Range: X/R	13.6 and 16.5 (Note 2)	2.1 - 18.8	3.5 - 20	6 - 30
Terminal, Typical: X/R	NA	8.2	15	NA
Line, Range: X/R	NA	NA	NA	1.6 - 3.1 (Note 3)
Line, Typical: X/R	NA	NA	NA	2.5 (Note 3)

NOTE:

1. X/R at the terminals specified in the subtransient condition.

2. Only two terminals.

3. Estimated from the distribution of X/R.

185

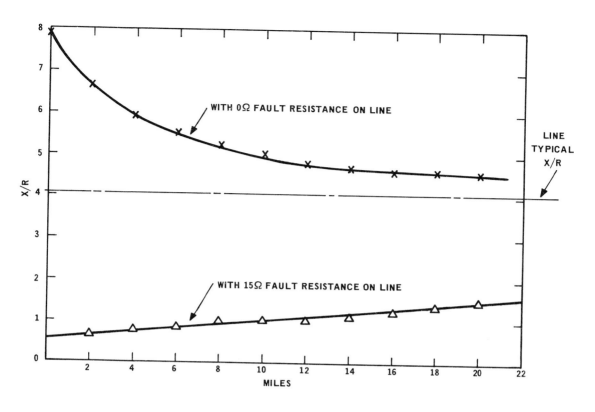

FIGURE 3-63

Typical X/R Variation With Separation From Terminal
for 110 kV Class Lines (Hydro Quebec)

Assuming these maximum figures:

$$X/R = 40 \rightarrow \rho = \arctan 40 = 88.6° \simeq \pi/2$$

$$R/L = \frac{\omega R}{\omega L} = \frac{\omega \cdot R}{X} = 377 \times \frac{1}{40} = 9.4$$

3.17.4 Resistive Coupling

Near the source, Equation (1) becomes

$$i = \frac{K}{Z} \{\cos(\omega t + \psi) - \cos \psi \, e^{-10t}\} \quad . \tag{3}$$

Because the time constant (1/10 s) is much greater than the half cycle time (1/120 s), the decay can be neglected during the first half cycle, and

$$i \simeq \frac{K}{Z} \{\cos(\omega t + \psi) - \cos \psi\} \quad . \tag{4}$$

The maximum current is reached at $t = \pi/\omega$, and its absolute value is then

$$i = \frac{2K}{Z} |\cos \psi| \quad .$$

Thus, with $\psi = 0$, the current can reach twice its steady state value at time $t = \pi/\omega$.

Neutralizing transformers installed to protect telecommunication circuits serving power stations have one of their windings directly connected to the neutral of the station. If the current contains a high dc component, the core may saturate, rendering the transformer ineffective. When neutralizing transformers are tested at substations, the short circuit applied should correspond to a minimum voltage in order to emphasize the overshoot and offset phenomena.[41]

Conditions resulting in saturation will be produced by circuits having long time constants (i.e., large X/R values). The data available show that the largest time constant occurs for faults close to source and is about 0.1 second whereas at about 20 miles from the source, the time constant is 0.056 second.

3.17.5 Distribution Circuits

Although limited, existing data on X/R ratios of distribution circuits indicate that the following figures are typical:

Maximum X/R at source: 10

Maximum X/R on line: 4

These values give the following results:

ρ source = arctan 10 $\simeq$ 84°

ρ line = arctan 4 = 76°

Time constant, source $\simeq$ 0.025 s

Time constant, line   = 0.01 s

If at closure, ψ = 0, the following results are obtained from equation 1.

Time at which the current reaches maximum at the source:

$$\frac{31}{32} \cdot \frac{\pi}{\omega} \simeq \frac{\pi}{\omega} \, .$$

Corresponding maximum value at the source:

1.72 $\times$ permanent fault value.

Time the current reaches a maximum on the line is given by

$$\frac{29}{30} \cdot \frac{\pi}{\omega} \simeq \frac{\pi}{\omega} \, .$$

Corresponding maximum value on the line:

1.44 $\times$ permanent fault value.

Figure 3-64 shows the variation of transient current with time.

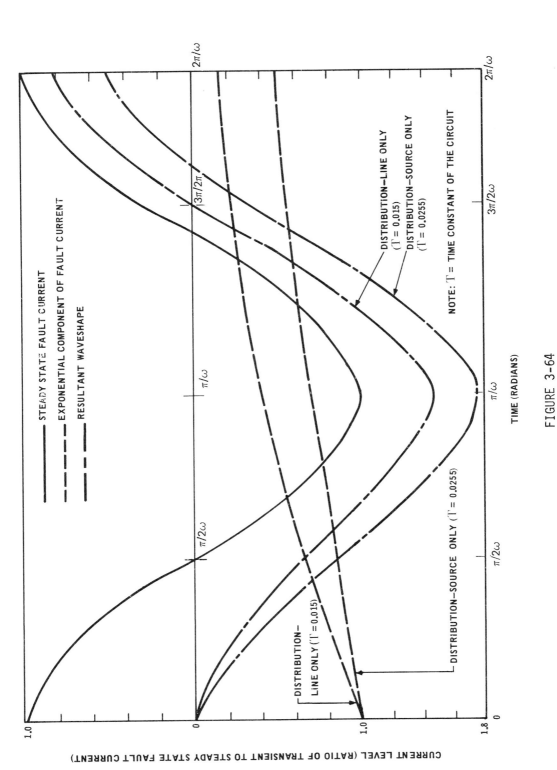

FIGURE 3-64

Variation of Transient Current Per Unit of Steady State Current With Time

3.17.6 Variation in the Reactances of the Power Network

In the preceding data, it has been assumed that the fault impedance was constant with time. The fault current expressed in Equation
(1) is a function of $Z_0 + Z_1 + Z_2$ (the zero, positive, and negative
sequence impedances from the point of fault). The zero and negative
impedances are constant, but the positive impedance in addition to a
constant component (e.g., lines, transformers, and the load) has a
component that varies with time (e.g., turbine generator, salient
pole generator, synchronous condenser, and other rotating equipment).

For a short circuit in a generator equipped with damper windings, or in a turbine generator, the armature current is made up of
an aperiodic component generated by the inductor and the damper
windings, and a constant-amplitude component corresponding to the
'permanent' short circuit.[8]

Figures 3-65 and 3-66 show the variation of short-circuit
currents in rotating equipment.

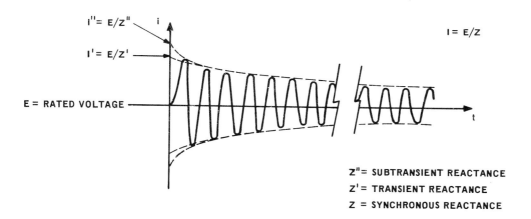

FIGURE 3-65

Variation of the Phase Current of a Generator
Under Short Circuit Condition

The short-circuit current (I_{SC}) is of the form

$$I_{SC} = [(I''-I')e^{-\alpha''t} + (I'-I)e^{-\alpha't} + I] \cos(\omega t + \psi).$$

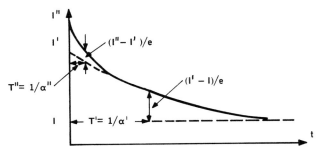

FIGURE 3-66

Amplitude Envelope of Armature Current[4]

The positive sequence impedance thus varies from a subtransient low, through a transient medium, to a steady state high value of synchronous reactance.

TABLE 3-29

Typical Values of Reactances and Time Constants
(After Reference 8)

	SUBTRANSIENT	TRANSIENT	SYNCH.	T''	T'
	Z''(%)	Z'(%)	Z(%)	(Hz)	(Hz)
Turbine Generators	10-25	15-30	120-250	1-3	60-180
Salient Pole Generators	13-35	20-45	60-125	2-6	60-180
Synchronous Condensers	25-30	40-50	100-200	1-6	60-180

The values of fault current used by power authorities in protection consideration are usually estimated on the basis of transient reactance. This approach simplifies calculations and is representative of the energy dissipated. In most cases, the effect of the variation of positive reactances is masked by the line.

3.17.7 Example

Assume a 3-phase neutral-grounded steel-tower 132 kV line as shown in Figures 3-67 and 3-68 and assume that all resistances can be neglected. All percentage reactance ratings for the components in Figure 3-67 are based on the corresponding MVA rating. Selecting a base of 50 MVA for the overall system (i.e., for each component), the results shown in Figures 3-69 through 3-71 are obtained.

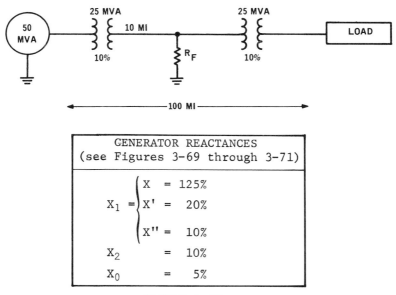

GENERATOR REACTANCES (see Figures 3-69 through 3-71)	
$X_1 = \begin{cases} X = 125\% \\ X' = 20\% \\ X'' = 10\% \end{cases}$	
$X_2 = 10\%$	
$X_0 = 5\%$	

FIGURE 3-67

Example: Line Configuration

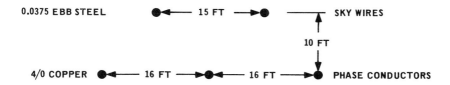

FIGURE 3-68

Example: Line Geometry

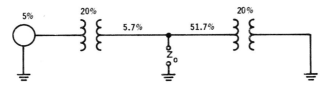

FIGURE 3-69

Zero Sequence Circuit (Results)

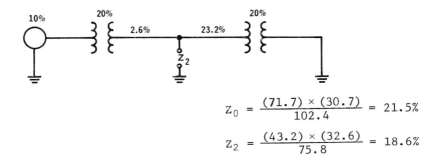

$$Z_0 = \frac{(71.7) \times (30.7)}{102.4} = 21.5\%$$

$$Z_2 = \frac{(43.2) \times (32.6)}{75.8} = 18.6\%$$

FIGURE 3-70

Negative Sequence Circuit (Results)

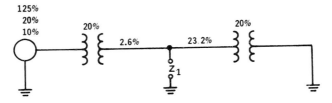

FIGURE 3-71

Positive Sequence Circuit (Results)

From Figure 3-71, Z_1, Z_1', and Z_1'' can be found as follows:

With the synchronous reactance:

$$Z_1 = \frac{(43.2) \times (147.6)}{190.8} = 33.4\% \quad .$$

With the transient reactance:

$$Z_1' = \frac{(43.2) \times (42.6)}{85.8} = 21.4\% \quad .$$

With the subtransient reactance:

$$Z_1'' = \frac{(43.2) \times (32.6)}{75.8} = 18.6\% \quad .$$

Using the synchronous reactance,

$$\textstyle\sum Z = 21.5 + 18.6 + 33.4 = 73.5\% \quad .$$

Using the transient reactance,

$$\textstyle\sum Z = 21.5 + 18.6 + 21.4 = 61.5\% \quad .$$

Using the subtransient reactance,

$$\textstyle\sum Z = 21.5 + 18.6 + 18.6 = 58.7\% \quad .$$

$$\text{Ratio } \frac{\Sigma Z \text{ sync.}}{\Sigma Z \text{ sub.}} = 1.25$$

$$\text{Ratio } \frac{\Sigma Z \text{ trans.}}{\Sigma Z \text{ sub.}} = 1.05$$

This example shows that even for short circuits close to the generating site, the ratio of maximum fault current computed with the transient instead of the subtransient reactance yields currents that are smaller by only 5 percent. If the maximum residual fault current is 4000 A, there will be a maximum increase of 200 A for a maximum period of 5 cycles.

3.18 CUMULATIVE DISTRIBUTION CURVES OF INDUCED VOLTAGES - APPLICATION TO FUTURE SYSTEMS

3.18.1 General

The levels, durations, and yearly occurrences of induced voltages on a system along a given right-of-way depend on:

a) the shielding of the system,

b) the density of exposures along the route,

c) the voltage classes of the neighboring power lines and the fault position with respect to the power terminals,

d) the coupling impedance of each exposure, and

e) the average distance between protectors and their breakdown value.

The relative positions between communication and power circuits, the density of exposures, and the voltage classes are all right-of-way (R/W) constraints which influence protection requirements and thus must be considered in the selection of a new telephone route. Spacings between protectors also affect the levels of voltage stresses and the magnitudes of resulting currents. Figures 3-72 and 3-73 illustrate the variations of voltage stresses due to variations in spacing between protectors or grounding points.

Predictions of future changes in inductive exposures depend on a number of parameters, which may or may not vary simultaneously. The best approach is to consider the existing network situation, and assess the effect of varying major parameters individually.

In existing transmission systems or those planned for the near future, the distances between protector points range from 1 mile (for an LD-1 system) to 120 miles (for the long haul LD-4).

Assuming that the right-of-way (R/W) parameter is fixed (i.e., future cable routes will be identical to the existing ones) and that neighboring power lines will not vary in position or voltage class, one can then 'test' the hypothetical present and future inductive interference situation for several typical configurations as follows:

a) the type of system considered - long, medium, short haul, or distribution,

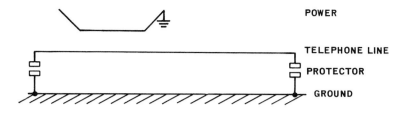

POWER

TELEPHONE LINE

PROTECTOR

GROUND

(A) Exposure Between Protector Points

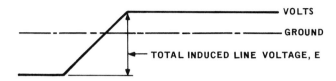

VOLTS

GROUND

TOTAL INDUCED LINE VOLTAGE, E

(B) Volts to Ground on Telephone Line Distributed
According to Impedance to Ground (Protectors have
not Operated)

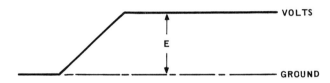

VOLTS

E

GROUND

(C) Volts to Ground (Left Protector Operated)

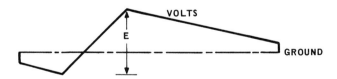

VOLTS

E

GROUND

(D) Volts to Ground (Both Protectors Operated, Grounding
Resistances Equal but Different from Zero)

FIGURE 3-72

Distribution of Longitudinal Voltage
(Single Feed Power)

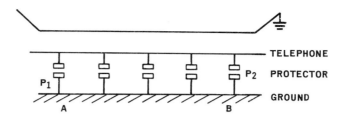

(E) Exposure Covering Several Protector Sections

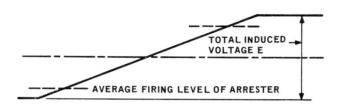

(F) Volts to Ground on Telephone Line Before any Protector Operation

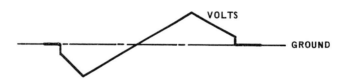

(G) Protectors P_1 and P_2 Operate and Reduce Voltage As Indicated

FIGURE 3-72 (Cont'd)

Effects of Protector Operation on Induced Voltage (Single Feed Power)

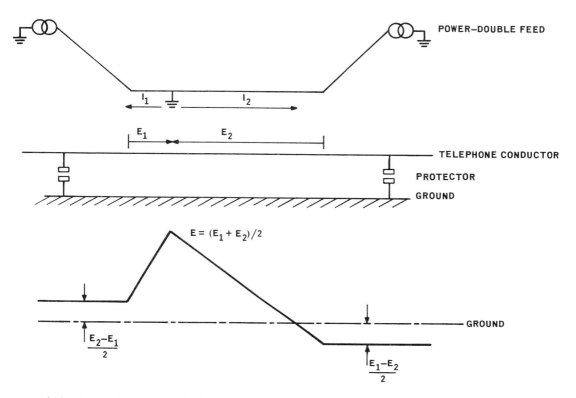

(A) Distribution of Voltages to Ground Prior to Protector Operation

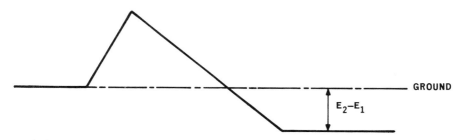

(B) Voltages to Ground (Left Protector has Fired)
Note: If $E_1 = E_2$, protectors will not fire.

FIGURE 3-73

Effect of a Double-Feed Exposure
on Voltage Stress

b) the terminal points, and

c) the media of transmission - paired cable, coaxial cable,
    or open wire.

This approach has the following merits:

1) the system requirements will be based on near-realistic
    cases,

2) the requirements will, in turn, set a pattern in the
    right of way selection, and

3) the chances of setting up interference standards with
    power utilities will be improved.

Among the important statistics that systems designers and
maintenance engineers require is the frequency of occurrence of
various levels of induced voltage. Distributions of this parameter
will vary according to spacings between repeaters. A method of
determining the distributions is given below.

3.18.2 Algorithms

A. ASSUMPTIONS

A.1 All contributing power sections have a uniform rate of
     faults.

A.2 The cumulative coupling is a linear function of the
     distance along the exposure; i.e., the inducing power line
     is parallel to the telephone line through the equivalent
     separation.

A.3 The distribution of residual current along the length of
     the power lines is of the form $K/(Ax+B)$, where x is the
     distance from a reference terminal.

A.4 The resistances of the terminal stations are negligible.
     The resistance at the point of fault has value R.

A.5 The telecommunication line is continuous over the
     whole length of the route.

A.6 The faults are uniformly distributed over the entire
     length of a power section.

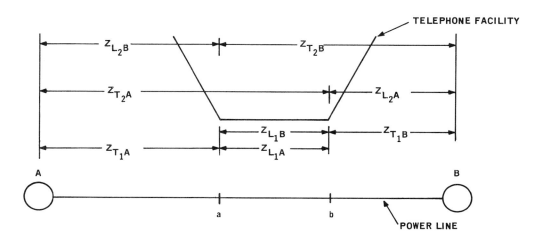

FIGURE 3-74

Symbols for The Derivation of Algorithms

The impedances in Figure 3-74 are in ohms, and the distances are in miles.

Let

Z_C = coupling impedance per unit length (Ω/mi),

ab = exposure length (mi),

aA, bB = distances from extremities of exposure to nearest power terminal point (mi).

B.   SINGLE EXPOSURE

B.1   Single Feed

Let

A = feed point,

B = load.

Consider first faults on ab. With the origin at a, the induced voltage for a fault at x within the exposure is given by

$$V_i = \frac{K}{Z_{T_1A} + Z_{L_1A} \cdot \frac{x}{ab}} \cdot Z_c \, x \, .$$

V_i is a maximum when $x = ab$, i.e.,

$$V_{maxA} = \frac{K}{Z_{T_1A} + Z_{L_1A}} \cdot Z_c \cdot ab .$$

Probability

$$P \, (V_i/V_{maxA} > \chi_i) = P_c \, (\frac{x}{ab} \geqslant \frac{\chi_i}{\{1 + (Z_{L1A}/Z_{T1A})\}(1 - \chi_i)})$$

$$= \int_a^1 p \, (\rho) \, d\rho$$

where

$$a = \frac{\chi_i}{\{1 + (Z_{L1A}/Z_{T1A})\}(1 - \chi_i)} ,$$

$p(\rho)$ = probability density of the position of the fault.

Because of assumption A.6 ($p(\rho) = 1$),

$$P_{ab} \, (V_i/V_{maxA} > \chi_i) = \frac{\{1 + (Z_{L1A}/Z_{T1A})\}(1 - \chi_i) - \chi_i}{\{1 + (Z_{L1A}/Z_{T1A})\}(1 - \chi_i)} \cdot \quad (1)$$

Consider now faults on bB, i.e., faults beyond the exposure, with the origin at b:

$$P_{bB} \, (V_i/V_{maxA} > \chi_i) = \frac{(1 - \chi_i)}{\chi_i (Z_{L2A}/Z_{T2A})} \cdot \quad (2)$$

(Faults on aA will not yield any induced voltage.)

Let

R = relative frequency of occurrence of L/G faults per 100 mi per year for the power line being considered.

The number of times per year that the induced voltage will equal or exceed the value $V_{maxA} \cdot \chi_i$ is

$$N_i = [(P_{ab} \cdot ab) + (P_{bB} \cdot bB)](R/100) \quad . \tag{3}$$

B.2 Double Feed

Double feed is applied on the majority of higher voltage circuits and on some 100 kV circuits. At the start of the fault, the two sides feed opposite currents, resulting in opposite induced voltages. There will be a period, however, when only one side is feeding and the other is open circuit. (This is due to slight time differences in the energizing times of the relays at each end.) To take this effect into account, and to allow for the fact that faults within the exposure could be fed from either direction, a weighting is applied to faults within ab.

The two feeding directions are usually asymmetrical both physically and electrically. Therefore, both directions are referenced to a common value, e.g., V_{maxA}. The resulting yearly occurrence of an induced voltage equal to or exceeding $V_{maxA} \cdot \chi_i$ is given by

$$N_i = \left\{ \frac{1 - \chi_i}{(Z_{L2A}/Z_{T2A})\chi_i} \cdot bB + \frac{(1 - \chi_i)(V_{maxA}/V_{maxB})}{(Z_{L2B}/Z_{T2B})(V_{maxA}/V_{maxB}) \cdot \chi_i} \cdot aA \right.$$
$$\left. + \left[\frac{\{1 + (Z_{L1A}/Z_{T1A})\}(1 - \chi_i)}{1 + \{(Z_{L1A}/Z_{T1A})(1 - \chi_i)\}} + \frac{\{1 + (Z_{L1B}/Z_{T1B})\}[1 - \{\chi_i(V_{maxA}/V_{maxB})\}]}{1 + [(Z_{L1B}/Z_{T1B})\{1 - \chi_i(V_{maxA}/V_{maxB})\}]} \right] \frac{ab}{2} \right\} \frac{R}{100} \tag{4}$$

C. THE PRESENCE OF OTHER POWER LINES

This is the case where more than one power line affects the telephone route (Figure 3-75). Because there is only a remote possibility that two lines simultaneously experience a fault, it can be assumed that the presence of other lines affects only the number of disturbances on the telecommunication line and not significantly the levels of those disturbances.

Most of the L/G faults originate from lightning hits on sky-wires so the adjacency of power lines can reduce the total number of yearly disturbances that might occur.

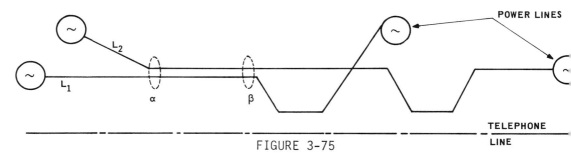

FIGURE 3-75

Power Lines Sharing the Same Right-of-Way

If lines L_1 and L_2 of the same voltage class share the same
right-of-way between α and β (see Figure 3-75), the effective
yearly L/G outage over $\alpha\beta$ of each will be R/2, where R is the
class outage. If the lines are separated by more than three times
the average height of the conductors, there will be no reduction.
If the lines within the shielding separation from each other are
of different voltage classes, say 735 and 110 kV, the higher voltage
class line will shield the other line almost totally over the
length $\alpha\beta$.

D. EFFECTIVE SECTIONALIZATION OF THE TELECOMMUNICATION LINE

Because telecommunications lines are sectionalized by power
feed points from a longitudinal induction standpoint, only a small
portion of the telephone route should be considered (see Figure 3-76).
The longest power feed sections of existing systems are:

N Carrier:	25 miles
Tl Carrier:	7 miles
LD1 Carrier:	12 miles
LD4 Carrier:	120 miles

The annual frequency of operation of a protector depends on
its firing level and on the spacing between protector points.

The longest spacings between protector points (repeater spacings)
are:

N Carrier:	8.4 miles
Tl Carrier:	1.14 miles
LD1 Carrier:	1.14 miles
LD4 Carrier:	120 miles (Coaxials)

Consider the exposures shown in Figure 3-76. The voltages
induced on sections outside the power section (i.e., on a4b4) will
not affect the power section considered. From Figure 3-76 it is

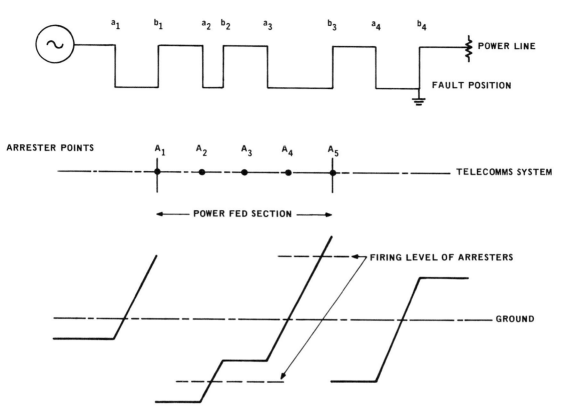

(A) Voltage to Ground Before any of the Protectors
A_5, A_3, A_2 or A_1 have Operated

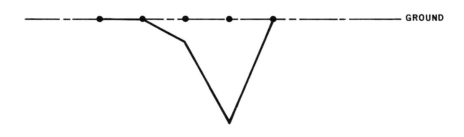

(B) Voltages to Ground After A_1, A_2, A_5 have Operated

FIGURE 3-76

Sectionalization of Exposures

also clear that the maximum current through a protector will be that produced by the worst exposure over a protector section length. From the point of view of the frequency of protector operation, only the power feed section should be considered.

3.19   RECOMMENDATIONS

3.19.1 Pilot Studies

Most of the results presented in this section have been theoretically derived. It is therefore recommended that two pilot circuits be selected, one in Hydro Quebec territory and the other in Ontario Hydro territory and extensive studies be made to verify or improve the present results. Such a field study would also suggest the degree of correlation to be expected between future predicted trends and actual future problems.

3.19.2 Earth Resistivity

The earth resistivity parameter plays a very important role in power interference. Very large assumptions were made to obtain the figures on earth resistivity in this section. It is therefore recommended that more accurate information be obtained by actual field measurements, probably in cooperation with other studies such as geological surveys, etc.

3.19.3 Arrangements With Power Companies

The need for coordination between the communications and power industry is today greater than ever in view of the foreseen expansion of the power industry into UHV systems and the modernization of the expanding communication industry with more sensitive components. Although the power companies are usually cooperative in providing requested data, the actual transfer of information often takes a long time. Without better planning and coordination, current statistical data and up-to-date information are likely to remain difficult or even impossible to obtain.

It is recommended that a liaison group be created to facilitate a smooth and rapid exchange of information between the two utilities in the research and planning stage. The basic idea is to permit appreciation of each other's problems with mutual understanding so that optimal technical solutions can be acheived. It is hoped that such a group would result in the solution of many potential interference problems in the planning stage.

3.19.4 A Word of Caution

The results presented in this section are based exclusively on Hydro Quebec and Ontario Hydro networks and consequently the results should not be generalized to other power systems which may have very different characteristics.

3.19.5 Continuous Review

The electrical environment is continuously changing, especially in the area of power utilities where radical changes are expected during the next decade as a result of advances in nuclear power generation accompanied with possible high voltage dc transmission and automated distribution. Because of such changes and the inevitable change in the communication facilities themselves, it is recommended that the basic information in this report should be continuously up-dated and revised to keep up with changing demands in the future.

3.20 REFERENCES

1. Electric Utility Engineering Reference Book, Vol. 3, "Distribution Systems", *Westinghouse Electric Corp.*, 1965.

2. EHV Transmission Line Reference Book, *Edison Electric Institute*, 1968, Chapters 3 and 4.

3. Electrical Transmission and Distribution Reference Book, *Westinghouse Electric Corp.*, 1965.

4. "Teleprotection", *CIGRE*, 1970.

5. cf Reference 3, Chapters 11 and 17.

6. cf Reference 1, "Protection Device Coordination" (Chapter 10).

7. Engineering Design Information for Power Industry Communication Services, Section 4, *AT&T*, 1971.

8. "Electric Power in Canada-1970-Energy Development", *Dept. of Energy, Mines, and Resources*, 1970.

9. Statistical Yearbook, *Ontario Hydro*, 1971.

10. Hydro Quebec Rapport Annuel, *Hydro Quebec*, 1971.

11. Carson, *BSTJ*, October 1926.

12. *CSA*, Standard C22.3, No. 1 - 1970.

13. BSP 876-102-100, *Bell System Practices*.

14. BSP 876-103-100, *Bell System Practices*.

15. BSP 876-420-100, *Bell System Practices*.

16. Discussion with A. Gahir, Lachine Cable Laboratory, 1973.

17. Discussion with R.S. Kallmeyer, Central Area, 1973.

18. Bell System, *Edison Electric Institute* Engineering Reports.

19. H. Peck, "La Protection des Cables à Enveloppe d'acier Contre les Tensions Induites Dangereuses", *Cables et Transmission,* 20^{e}A, No. 3, 1966.

20. McKenna, "Induced Voltages in Coaxial Cables and Telephone Lines", *CIGRE* Conference, Paper #36-01, 1970.

21. Sunde, "Earth Conduction Effects on Transmission Systems," 1968, Dover Publications.

22. L.G. Posdnjykov, "Transactions of the Osmk Institutes of Railway Engineering, Vol. 52, 1965, p. 99 (Referenced in *CIGRE* Conference 1970, Paper #36-01).

23. *AIEE* Transactions, Vol. 37, Part II, 1918.

24. Jean Fallou, "Lecons d'Electrotechnique," Chapter 12, Gauthier-Villars, Paris, 1949.

25. Rockefeller, "Ground Distance Relay," *IEEE* Transactions on PAS Vol. PAS 85 #10, October 1966, p. 1027.

26. Hydro Quebec Report, "Mesure de la Resistance Pylone Terre des Circuits 1313 et 1312, Pandora - Rouyn", October 1968.

27. *IEEE* Committee Report, "A Guide for the Protection of Wire Line Communication Facilities Serving Electric Power Stations", PAS 85, #10, October 1966.

28. Knowlton, "Electrical Engineering Handbook", Section 24-29, McGraw - Hill, 1970.

29. C.F. Minutes of Meeting Held With Hydro Quebec Representatives on March 13, 1973 (Record kept on file #7.3.1.1 in Dept. 3E00).

30. *CCITT* Directives 1963, p. 55.

31. Bell *EEI* Engineering Report #39, paragraphs 2 and 6, p.p.17.

32. B.S.P., AB64.329, *Bell System Practice*, p.p.18.

33. *IEEE Transactions*, Vol. PAS-89, Nov/Dec 1970, p.p.1893.

34. *Australian Communication Monographs*, No. 3, Nov. 1964, p.p.29.

35. M. Syed, "The Effects of Corona Discharge on Communication Networks,"Technical Report, *Bell-Northern Research*, Ottawa, 1973.

36. cf Reference 30, Chapter XI.

37. Electrical Transmission and Distribution Reference Book, *Westinghouse Electric Co.*, 1950.

38. BSP, AB63.232, *Bell System Practice*.

39. cf Reference 21, Paragraphs 7.2 to 7.7.

40. *Bell Canada* Headquarters Design Memorandum, 71-02, April 1971.

41. Bell EEI Engineering Report #44.

4. ELECTRIC SHOCK

4.1 INTRODUCTION

Electric shock from power facilities is most frequently the result of direct contact with an energized conductor. However, a direct contact is not necessarily required. For example, electric shock may also result from the following types of exposure:

1) Leakage currents through insulation.

2) Flashover along surfaces.

3) Discharge through air as a result of close proximity to a high voltage conductor.

In modern society, man is seldom unexposed to the possibility of electric shock. However, due to a concerned safety effort, cases of fatal electric shock have been kept small in comparison to fatalities from many other common hazards. Statistics Canada data on the annual number of deaths in Canada from electrocution by power frequency currents for the years 1967 through 1971 are given in Table 4-1. No reliable information is available on the number of nonfatal shocks, but it seems reasonable to assume that they occur at a much higher rate than fatal shocks. Such nonfatal shocks cannot be ignored, since the associated muscular reaction may trigger events that lead to physical injury from falls, spilling of hot liquids, etc. The 'surprise factor' responsible for such events, which may occur as the result of even a modest shock, requires consideration in assessing probable hazards. To complete the shock statistics, the reported deaths in Canada from lightning are given in Table 4-2. One would expect that the number of lightning incidents would be strongly influenced by the incidence of thunderstorm days, the proportion of rural population, and rural safety habits.

This relatively good record *should not breed complacency*, because changes in electrical practices can create new aspects to the problem of safety. It is essential that those who design electrically powered equipment and develop industrial procedures understand the nature of electric shock and be familiar with the magnitudes of electric current that produce adverse effects on the human body in order that they may meet their responsibility for safety. Practical obstacles to absolute electrical safety still exist, and basically the problem remains one of probability, involving several diverse variables.

The consideration of electric shock involves two major aspects.

1) Physiological effects associated with current flow through the body. This involves such aspects as the electrical equivalent of the human body, degrees of shock and related physical manifestations, response to different duration or severity of shock.

2) Evaluating shock probabilities, developing safety practices, and establishing safety levels.

Item 1, dealing with technical aspects of electric shock, will be covered in this section. Item 2, involving the determination of acceptable levels of risk for specific applications, will not be discussed here, since the subject cannot be satisfactorily generalized and sometimes involves matters of policy. The following statement summarizes some of the considerations regarding the establishment of acceptable levels of risk.

Safety is to some extent a philosophical subject. Although one would appreciate having a formula applicable to all situations to reduce the responsibility of personal judgement, no such simple procedure is available. Safety decisions involve such practical considerations as technical feasibility and economics, and there is no established safety level applicable to all situations. In some cases the nature of an activity may be so rewarding that relatively high risks are generally accepted. Consider, for example, the case of automobile transportation.

The probable risk level of electric shock for a specific situation can be estimated with the information to be presented herein but the decision as to the acceptability of this risk remains a matter of judgement.

In the 1920's, L.P. Ferris *et al*[1] attempted to obtain definitive information applicable to humans concerning the hazardous shock range. Several investigators subsequently contributed to this effort, their data based chiefly on measurements made on animals and observations on humans in the nonlethal range. These studies have been supplemented by investigation of accidental shocks suffered by humans. Information abstracted from these sources that is believed to have engineering application in the communication field is presented in this section.

TABLE 4-1

Reported Deaths in Canada From Electrocution by 60 Hz Currents

(Statistics Canada)

DATE	POPULATION (10^6)	TOTAL DEATHS		HOME WIRING AND APPLIANCES		INDUSTRIAL		MISCELLANEOUS		DEATHS PER MILLION OF POPULATION	
		MALE	FEMALE	MALE	FEMALE	MALE	FEMALE	MALE	FEMALE	MALE	FEMALE
1967	20.38	82	5	-	-	-	-	-	-	4.0	0.25
1968	20.70	53	0	-	-	-	-	-	-	2.5	0.0
1969	21.10	65	5	4	3	28	0	33	2	3.1	0.24
1970	21.30	65	3	7	1	20	1	38	1	3.1	0.14
1971	21.57	73	3	12	2	24	0	37	1	3.4	0.14
AVERAGE ANNUAL	21.0	67.6	3.2	7.7	2	24	3	36	1.3	3.2	0.15

TABLE 4-2

Reported Deaths in Canada From Lightning

(Statistics Canada)

DATE	POPULATION (10^6)	TOTAL DEATHS		DEATHS/MILLION OF POPULATION	
		MALE	FEMALE	MALE	FEMALE
1967	20.38	5	4	0.25	0.20
1968	20.70	7	2	0.34	0.10
1969	21.10	9	2	0.43	0.09
1970	21.30	14	1	0.67	0.05
1971	21.57	9	1	0.42	0.05
AVERAGE ANNUAL	21.0	8.5	2.0	0.41	0.10

4.2 DEGREES OF SHOCK

Starting from the threshold of perception, physical effects proceed as follows with increasing current through the body:

1) Mild sensation

2) Painful sensation

3) Inability to release if gripping an energized object

4) Respiratory paralysis

5) Ventricular fibrillation

6) Heart paralysis (no fibrillation)

7) Tissue burning

4.3 VARIABLES RELATING TO SHOCK SEVERITY

1) Type of current (steady state ac and dc, superimposed ac on dc, or impulse)

2) Magnitude of current

3) Frequency

4) Waveshape

5) Duration of shock

6) Physical condition of victim (weight, body structure, state of health, sex, age)

7) Current path through body

8) Phase of the heart cycle at the instant shock occurs

The contribution of each of the above variables will be subsequently evaluated on the basis of consensus among investigators. Where agreement does not exist, the range of opinion will be indicated.

4.4 PHYSIOLOGICAL EFFECTS

The following discussion of physiological effects is directly quoted from Kouwenhoven's paper *Effects of Electricity on the Human Body*, published in the March 1949 issue of Electrical Engineering:[2]

"The passage of an electric current through the body produces numerous effects that differ not only in intensity, but also in kind. They range all the way from a slight tingling sensation to death. The consequences depend upon the value, frequency, and pathway of the current and on the duration of the shock. The aftermath may be good or evil. An electric shock may produce healing in certain mental diseases or it may produce a state of depression of the vital processes of the body characterized by rapid but weak pulse, rapid but shallow breathing, pallor, restlessness, and a depressed mental state similar to surgical shock or a highly excited, almost maniacal state.

Some of the effects produced by an electric current are discussed in the following.

Conscious Phenomena. If the victim of an electric shock retains consciousness during and following the contact, there is often a whistling or ringing in the ears and partial deafness for a time. In addition, there may be visual disorders such as flashes and brilliant luminous spots. Pain and soreness of the muscles are a common reaction. If the shock is a severe one, the victim usually will be restless and irritable. These disorders generally disappear in a few hours.

Muscular contractions are produced when contact is made with an electric circuit. These contractions are particularly marked when the circuit is an alternating one of commercial frequencies. At high voltage the contraction of the muscles is very sudden and severe. It may throw the victim clear of the circuit. In some instances bones have been broken. The severity of the contraction probably accounts for the soreness that is felt in the muscles. Clonic contractions of the extremities often are observed following a shock and tremors may continue for some minutes.

Convulsions may occur in cases of electric shock. They usually are characterized by irregular muscular spasms and tremors....

Narcosis, a state of stupor and insensibility, has been claimed as one of the results of shock. When the passage of an electric current through the brain of a patient causes convulsions, it also produces unconsciousness lasting for several minutes. This unconscious period might be considered anesthesia in as much as the patient is insensitive to pain. Recently, this unconscious period has been employed by some surgeons to perform short operations, and others have attempted to produce periods of narcosis in the treatment of psychotic patients by utilizing lower values of electric current and thereby eliminating the convulsions.

Anesthesia, resulting from the passing of interrupted direct currents through the head, was claimed by Leduc of Leipzig in 1902. Other investigators, including the author, have repeated Leduc's experiments without success. Hertz, in his thorough study of the subject, found that anesthesia was not attained either in animals or man. Instead the patient would be rendered unconscious, and breathing would cease unless the strength of the shock was limited carefully. A so-called electric anesthesia is used by the British in the slaughtering of hogs. One electrode is placed in the mouth, the other on top of the head, and a 50-cycle current is passed directly through the brain, rendering the animal unconscious. The bleeding of the animal is reported to be better following this treatment than in normal slaughtering practice because of the increased blood pressure that follows the shock.

Loss of consciousness occurs in many electrical accidents. Sometimes the victim recovers spontaneously; in other cases, either after the application of artificial respiration, or never. Cases also have been reported where the victims lost consciousness when contact with the circuit was made at two points on the same leg or hand, and in which there was no burning of the tissues. Such cases are believed to be due to a severe shock to the system.

Electric burns are of two types: those produced by the heat of the arc, as may result when contact is made with a high-voltage circuit; and those that are caused by the passage of an electric current through the skin and the tissues. Burns resulting from an electric arc are, in general, similar to those produced by high-intensity heat sources. The true electric burn often is characterized by a pinkish mark on the surface of the skin. The burns, however, may penetrate deeply and require considerable time to heal. Jellinik reports a case where the current value was large enough actually to char the flesh at the elbow where there exists only a relatively small amount of body tissue. Burns, blisters, and markings are not necessarily present on the skin after an electrical accident. When the skin is saturated thoroughly with water and the contact area is not restricted, a fatal shock may not leave the slightest detectable blemish. Burns produced by electricity usually heal without infection. They, however, heal slowly. In severe cases, fingers or limbs may be lost and death may follow as a secondary effect.

Emission of seminal fluid is common in males and occurs at low as well as high voltages. It is believed that the severe muscular contraction is responsible for this effect.

Priapism has been observed in a number of instances and in some cases has continued for a week. This phenomenon is believed to be caused by an irritation produced in the spinal cord by the flow of current.

Incontinence is reported in some cases following a shock and in some instances blood is present in the urine.

Blood pressure rises suddenly when contact is made with a high-voltage circuit because of the severe muscular contraction. When the circuit is broken, the pressure usually remains high for a period due to the rapid heart rate; provided that organ is not injured by the shock.

Hemorrhages usually petechial in nature sometimes are found in the brain, the nervous system, and other organs. One or both eyes may be bloodshot following a shock, due to the rupturing of blood vessels in the conjunctiva. Gross hemorrhages have been found in the fourth ventricle of the brain upon autopsy.

The nervous system may be so profoundly shocked or fatigued by a contact with an electric circuit that it cannot function normally again for a period of minutes or hours. The nerve cells are injured, especially in areas that have been traversed by the current. Injured cells are characterized by a dark shrunken nucleus, which is often eccentric in position, and the loss of granules. The damage, however, is patchy in distribution so that injured and normal healthy cells lie in close proximity. Autopsy of shock victims also has revealed cavities in the nervous system of 25 to 200 microns in diameter. These may be caused either by heat or electrolysis.

One of the most common effects on the nervous system is the production of a temporary paralysis or block. The location of this block will depend upon the path taken by the current. The lungs or other portions of the body may be paralyzed following the shock. There is a case on record where a woman stood with her back resting against the edge of an electric range when the power line was struck by lightning. She received a severe shock which was followed by a temporary paralysis and loss of sensation in both limbs that lasted for about four hours. The many successful resuscitations resulting from the prompt application of artificial respiration to shock victims may be ascribed to the temporary nature of this paralysis. If nature is given the opportunity, it often will repair the damage and again permit the signal from the brain to reach the organ in question.

The damage that electricity produces in the nervous system is not specific in that other diseases give rise to similar patterns.

Ventricular fibrillation results when a small current passes through the heart and disturbs its normal co-ordinated rhythm. The human heart does not recover spontaneously from ventricular fibrillation. While the heart is in this condition there is no circulation, and death will ensue.

Ventricular fibrillation may be arrested by the passage of a 60-cycle current of the order of one to two amperes through the heart. This value of current is sufficient to bring the muscles of the heart to rest and hold that organ in diastole. Then when the circuit is broken the heart usually will resume its normal operating rhythm. The feasibility of this method of recovering the heart by an electric counter shock was demonstrated by using experimental animals. It has been applied to man and two cases of successful recovery of the fibrillating heart are reported.

Permanent Effects. Permanent injuries from contact with electric circuits fortunately are extremely rare. Perwitzschky reports 23 cases of auditory and vestibular injuries that appeared either immediately or from one to two years after the shock. It is peculiar that the damage was not related in any way either to the severity of the shock or to the path of the current through the body. There are cases on record where the ear formed one of the circuit contacts yet no permanent after-effects resulted.

Panse and others have reported the development of disseminated sclerosis following electric shocks. In all of these cases no pathological studies are reported, and for the present one must be skeptical of ascribing them to electricity.

Nerve lesions and paralysis are rare after electric shocks, but there are authentic examples. These usually are characterized by muscular atrophy in the extremity through which the current passed. Usually there was pain and weakness following the shock which gradually was transformed into a slow atrophy.

In assessing the value of reports on permanent injuries one should use considerable critical judgment. These cases are extremely difficult to evaluate because of the great desire for compensation. It is clear that most victims either die immediately or recover, perhaps after extensive burns have healed, without demonstrable neurological damage.

Death from electric shock may result from a number of causes or from a combination of two or more of them. *In general, low voltages kill through the mechanism of ventricular fibrillation and high voltages either through the destruction or inhibition of the nerve centers; asphyxia being the immediate cause of death.*" (our italics)

4.5 ELECTRICAL MODEL OF THE HUMAN BODY

The equivalent electrical circuit of the body consists of three parts (Figure 4-1):

1) where the current enters the body, the epidermis (skin) acts like a leaky capacitor;

2) the tissue of the body, which essentially presents a pure resistance; and

3) where the current leaves through the skin, which again acts like a leaky capacitor.

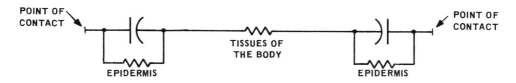

FIGURE 4-1

Electrical Equivalent of The Human Body

The resistance of the skin is not constant but varies with the amount of moisture that it contains, its temperature, and the applied voltage. Dry epidermis (outer skin) has a high resistance, which may reach 100 000 ohms per square centimeter.[2] The resistance offered by the body tissue is low, since, because of their salinity, fluids and blood are good conductors. The only poor conductors inside the body are bones; consequently the resistance of the body is relatively low.

Moisture is a major factor in the effective resistance of the skin, which drops substantially when wet. Measurements of human contact resistance for various types of skin-contact resistance (both dry and wet) obtained by R.H. Lee[3] are given in Table 4-3. Also, skin initially having a much higher resistance than the body is likely to lose its protection if it remains in contact with an energized surface for an appreciable time, because of the formation of blisters. At 50 V rms, blisters form in 6 to 7 seconds.[2] The results of body resistance tests conducted by C.F. Dalziel[13] are given in Table 4-4.

Dalziel[13] states:

"Skin resistance plays a very important role in limiting the current in low voltage accidents not only because of its effect in limiting the current directly but also because the low voltage may be insufficient to break down the protective resistance of the skin. In contrast, on high voltage circuits, sparks, arcs or corona may deteriorate the protection afforded by normal skin and may very rapidly produce deep burns and in addition tend to nullify the resistance of loose or imperfect connections and contacts."

TABLE 4-3

Human Resistance For Various Skin-Contact Conditions[3]

(Values indicate total human resistance - skin plus tissue)

CONDITION	RESISTANCE (OHMS)	
(AREA TO SUIT)	DRY	WET
Finger touch	40 k-1 M	4-15 k
Hand holding wire	15-50 k	3-6 k
Finger-thumb grasp*	10-30 k	2-5 k
Hand holding pliers	5-10 k	1-3 k
Palm touch	3-8 k	1-2 k
Hand around 1½-inch pipe (or drill handle)	1-3 k	0.5-1.5 k
Two hands around 1½-inch pipe	0.5-1.5 k	250-750
Hand immersed		200-500
Foot immersed		100-300
Human body, internal, excluding skin = 200 to 1000		

* Data interpolated.

TABLE 4-4

Body Resistance Tests*

SAMPLE SIZE	STEADY-STATE RESISTANCE IN OHMS		IMPULSE RESISTANCE IN OHMS AFTER 100-MILLIAMPERE IMPULSE	
	RIGHT HAND TO BOTH FEET	LEFT HAND TO BOTH FEET	RIGHT HAND TO BOTH FEET	LEFT HAND TO BOTH FEET
128	1260	1490	1190	1510
124	1630	1500	1390	1260
125	1450	1600	1410	1490
82	1370	1640	1300	1400
132	1250	1275	1200	1160
104	1480	1500	1710	1440
129	1460	1230	1135	1250
93	2150	1970	1730	1820
137	1650		1170	
Minimum Resistance	1230		1135	

* Contacts - Hand Grasping No. 7 Copper Wire and Both Feet
   Standing in Metal Pan Immersed in 3/4 Inch of Salt Water.

Dalziel does not define low and high voltage circuits in this con-
text, and such a statement is typical of a substantial amount of the
information in published literature on electric shock. It emphasizes
the inability and reluctance of even major investigators to commit
themselves to definitive statements without strong statistical
backing. This introduces difficulties in reducing information to a
form applicable to engineering problems.

A large number of measurements of human skin resistance were
made by H.B. Whitaker of the Underwriters Laboratories as part of
a study that evaluated the possible hazard of electric fences[15].
Interested readers are referred to Whitaker's paper.

The capacitance shown in the hypothetical model in Figure 4-1
is evident only at low currents. Kouwenhoven[2] demonstrated the
effects of capacitance by applying 50 volts from a current-limited
source to electrodes held in the hands. Immediately after closure,
a current of 19 microamperes was recorded, but after 500 microseconds,
the current had dropped to only 3 microamperes. From a more
practical standpoint, investigators generally agree that, in the
area of substantial shock at power frequencies, the contribution of
skin capacitance is negligible because of other factors that
drastically reduce the initial skin resistance. The generally
accepted model at power frequencies is purely resistive. The skin
may still contribute to total body resistance but is subject to
variables such as contact areas, moisture, tightness of contact,
cuts, and abrasions.

It is known that very small currents applied directly to the
heart can produce fibrillation; consequently the proportion of the
total shock current entering the body that reaches the heart is a
very critical factor. The distribution of current in the body is
influenced to a large extent by the locations on the body at which
contacts occur; thus the points where current enters and leaves
have a very important relationship to the degree of shock produced by any given body current. Body paths from arm to leg, which
produce a path diagonally across the chest cavity, and head to leg
are particularly critical. Arm to arm contacts are somewhat less
sensitive. The proportion of total shock current
reaching the heart from leg to leg contacts is relatively small.
Ferris[1] *et al* conclude that leg to leg contacts are unlikely to
produce fibrillation in humans, even at values of 15 amperes or
more. Such currents, however, would probably burn a victim unless
the contacts were good and the shock was of short duration.

In practice, there is very little control over the manner in
which a victim contacts an energized circuit. However, in
industrial accidents it has been noted that contacts from hand to
hand and hand to foot are common. Both result in relatively
hazardous body paths, which further suggests a conservative approach

in assessing hazards. It is also of practical interest to note that foot to foot contacts are much less critical, which is significant in judging the probable hazard for personnel walking in the area of earth potential gradients.

In developing a model for analytically predicting specific shock hazards, the degree of uncertainty regarding variables necessitates a conservative approach. Surprisingly there is very little information in the literature on modeling. Dalziel, in one of his papers[13], merely states that the commonly accepted value of body resistance between major extremities is 500 Ω. In a more recent paper[9], Dalziel states that:

> "A value of 500 ohms is commonly used as the minimum resistance of the human body between major extremities and this value is frequently used for estimating shock currents during industrial accidents. A value of 1500 ohms, which may be too high, is used to represent the body circuit between normal perspiring hands of a worker and in estimating currents of the reaction current level."

R.H. Lee[3], who is interested in industrial safety and has done some work in the area of body resistance, presents the values shown in Table 4-3 of this text. The values given are total body resistance for a variety of industrial conditions. The reader can draw his own conclusions where applicable. In non-industrial situations where the possible victims are likely to be adult females or children, it should be recognized that, in general, they have a substantially lower tolerance to electric shock than males.

In the Bell system, a value of 1500 ohms has frequently been used for estimating shock hazard.

4.6 BODY WEIGHT, CURRENT, AND DURATION

4.6.1 General

The type and magnitude of the current, the body weight of the victim, and the duration of exposure are very important factors in determining the degree of shock likely to be experienced. In communication systems, the possibility of accidental contacts and resultant body currents may be controlled to a practical degree by guarding, voltage limitation, and current limiting resistance. Duration is less readily anticipated or controlled. In some specific applications, however, a degree of time limitation is present that significantly mitigates the possible severity of shock, such as in the case of 20 Hz telephone ringing current, where the cycle is typically 2 seconds on and 4 seconds off. A body weight of 68 to 70 kg (about 150 lb) has been frequently employed for characterizing a normal man in generalized assessments of shock hazards. However, the recent trend appears to be towards greater conservatism in this respect. For example, in his papers[12,13] of 1946 and 1953, Dalziel uses a body weight of 70 kg in deriving his electrocution equation. In a more recent paper[5] (1968), co-authored by Dalziel and Lee, it is suggested that a body weight of 50 kg (about 110 lb) should be used in assessing human shock hazards. The interrelationship of these major parameters and their numerical evaluation with regard to the degree of shock produced will now be considered.

There is increasing evidence[5] that the most common mechanism of death from electric shock is ventricular fibrillation; consequently it has been a major area of investigation. This aspect of the problem should also be the primary concern of communications protection people and therefore will receive major consideration in this text.

4.6.2 Body Weight

Since the body weight of a victim is directly related to the degree of shock experienced, irrespective of the type of current involved, it will be discussed first.

Experimental studies have been chiefly confined to animals, of which the dog shows close correlation with man. Predicting the effects on man by scaling such data for the differences in weight introduces some degree of uncertainty. Dalziel and Lee have addressed this problem in considerable depth through a statistical correlation of their own data as well as that obtained by several other recognized investigators. They conclude that the minimum current required to produce ventricular fibrillation is approximately proportionate to an individual's body weight[5]. For practical purposes, this permits linear scaling. In earlier papers, body weights of 68 kg (150 lb) or 70 kg (154 lb) were frequently suggested as re-presentative values for estimating shock effects. On the basis of data obtained in Great Britain in a study of electrocutions (1962-1963), W.R. Lee[6] submits that a 'most' conservative value for characterizing a normally developed adult human should be substan-tially lower than values previously assumed. He proposes a weight of 50 kg (110 lb) for generalized considerations, and Dalziel and Lee[5], in their paper of 1968, base their derivation of an electrocution equation on that weight.

4.6.3 Threshold of Perception

Small currents, which merely produce an awareness of electric shock and produce no permanent adverse electrical effects on the body, may still constitute a safety hazard because of the possibility of involun-tary reactions that may lead to physical injury. The threshold of per-ception varies substantially between individuals and is typically in the range of about ½ to 2 milliamperes (Figure 4-2). The effect of frequency on perception current is shown in Figure 4-3, which is based on data obtained with subjects holding a small copper wire in their hand. The data indicate that the perception current at frequencies above that of commercial power is higher[9]. In general, leakage currents should be below the minimum level of perceptibility, and designers should be cognizant that there are codes stipulating maximum permissible leakage current values. Medical equipment has specialized maximum leakage current requirements which in some cases may be 10 microamperes or less (see National Electric Code, Section 517-520).

4.6.4 Let-Go Current

Let-go current is an important aspect of electric shock, because the inability to release oneself from an energized circuit can have serious consequences. Fortunately, investigation of let-go currents is relatively easy because data can be obtained directly on humans through controlled laboratory experiments with a high degree of safety. The subject has been relatively well investigated, and statistical data of a good degree of confidence have been published[9,10]. Figure 4-4, which is based on data obtained with subjects holding a small copper tube, gives the distribution of let-go current for both men and women.

An energized electrode held in the hand produces an increasing sensation of warmth (dc) or tingling (ac) as the current through the body is raised above the level of perception. As the current is increased, the effects become more severe. Muscular reaction increases and pain develops until a point is reached where the victim cannot voluntarily release the conductor. This is sometimes referred to as 'freezing' to the conductor.

Dalziel concludes, with respect to let-go current[9], that:

" ... the location of the indifferent electrode, moisture conditions at the point of contact, and size of electrode have no appreciable effect on an individual's let-go current."

He further concludes that:

" ... a normal individual can withstand, with no serious after effects, repeated exposure to his let-go current for at least the time required for him to let go."

Human response varies with current frequency as shown in Figure 4-5. For currents in the 25-60 Hz frequency range, response appears to be practically the same. These curves also indicate that let-go currents are higher at frequencies below and above the common power frequency range. The effects of current frequency are discussed further in the subsection on fibrillation.

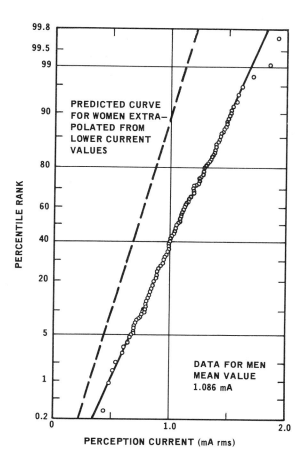

FIGURE 4-2

Threshold of Perception[9]

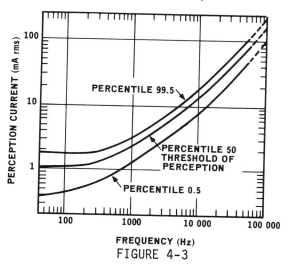

FIGURE 4-3

Effect of Frequency on Perception Current[9]

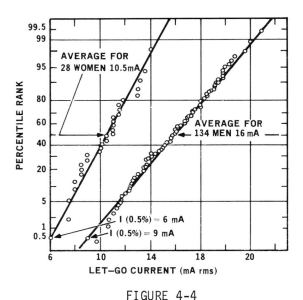

FIGURE 4-4

Let-Go Current for Men and Women[9]

Currents for men follow a normal distribution as do
those for women which use a smaller sample.

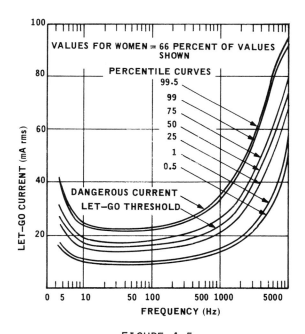

FIGURE 4-5

Let-Go Current Versus Frequency[9]

Currents become progressively more dangerous to an
increasing number of persons as shown by percentile values.

Currents Above Let-Go But Less Than That Producing Fibrillation

This is the area bounded by statistics on let-go and fibrillating currents. The probable results of such exposure are unpredictable, since shock severity increases with the duration of exposure and, once an individual has become frozen to a conductor, the condition may continue until the circuit is de-energized or muscular reaction produces physical deconnection.

Two investigators provide some information on the subject. Dalziel, in discussing effects on very high let-go currents[9], states:

" ... the muscular reactions caused by commercial frequency alternating currents in the upper ranges of let-go current, typically 18 to 22 mA or more, flowing across the chest cavity stopped breathing during the period the current flowed, and in several instances caused temporary paralysis of the middle finger. However, normal respiration resumed upon interruption of current, and no adverse effects were produced as a result of not breathing for short periods."

W.R. Lee[11], in a study of 30 cases of nonlethal shock in which the currents were above let-go values but below those producing fibrillation, found that, where persons were frozen to the circuits before becoming subsequently free, the victims who suffered the longer shocks showed signs of impending asphyxia during the periods they were held in contact with the circuit.

4.6.5 Fibrillation With Relation To Current And Shock Duration

50-60 Hz CURRENTS. Dalziel introduced the concept of an electrocution equation in 1946 with suggested values for estimating fibrillation thresholds[12]. Subsequently (1968), in the Dalziel and Lee paper[5], the authors retain this general approach but recommend more conservative values in the equation. This follows from generalizing human body weight at 50 kg rather than 70 kg. They present a method for estimating the probability of fibrillation with respect to two current levels: (1) minimum fibrillating current, and (2) definite fibrillating current. They also submit an expression valid for shock durations from 8.3 milliseconds to 5 seconds. With respect to the shorter time corresponding to a half wave of 60 Hz ac, it is stated that the majority of shorter shocks may be classified as impulse shocks. This latter type of shock, which appears to relate more closely to energy in the wave rather than current-time, will be covered in a subsequent section.

ELECTROCUTION EQUATION. The electrocution equation is

$$I = \frac{K}{\sqrt{T}} ,$$

(1)

where

I = Critical current in milliamperes,

K = A constant derived on the basis of body weight,

T = Duration of shock in seconds.

An appropriate value for the constant K can be derived from the curves in Figure 4-6. These curves were derived on the presumed response of the lower half percent of any normal human adult population selected on the basis of weight. The authors[5], in describing the basis for these curves, state that actual current producing fibrillation is greater than indicated by the lower line and in general will distribute somewhere in the area between the two lines. It seems certain, therefore, that values of K obtained through the use of the lower curve will produce conservative estimates.

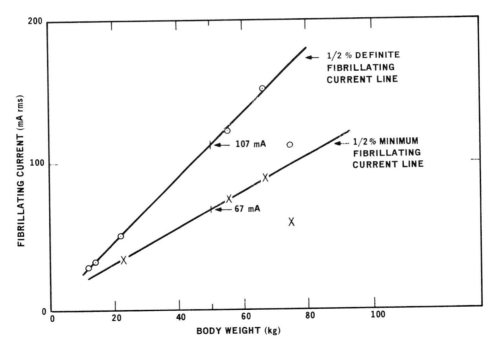

FIGURE 4-6

Relation of Fibrillating Current to Body Weight[5]
(3.00-second Shocks)

The use of expression (1) is demonstrated in the following example.

Conditions:

1) Body Weight = 50 kg,

2) Shock Duration = 2 seconds.

For minimum fibrillating current I_1, enter Figure 4-6 (based on 3 seconds shock duration) on the abscissa at 50 kg and obtain a corresponding current of 67 milliamperes from the lower curve. Then,

$$K = \sqrt{3} \times 67 = 116,$$

$$I_1 = \frac{116}{\sqrt{2}} = 82.3 \text{ mA}.$$

To obtain definite fibrillating current I_2, follow a similar procedure using the upper curve. Then,

$$K = \sqrt{3} \times 107,$$

$$I_2 = \frac{185}{\sqrt{2}} = 131 \text{ mA}.$$

The actual fibrillating current will therefore lie between 82.3 and 131 mA.

Assuming a severe contact situation where the body resistance may be 500 Ω, it appears from the above that voltages within the range of 41 to 65.5 V rms and applied for 2 seconds could possibly cause ventricular fibrillation in a 50-kg human. Kouwenhoven states in a 1949 paper[2] that several cases were already on record at that time where 60-65 volt circuits of common power frequency had produced fatal accidents. He further states that the lowest voltage fatality of which he had any record occurred at 46 volts, 60 Hz. Information such as this, in conjunction with subsequent experimental data, is probably what prompted the AT&T to reduce permissible *steady state ac voltages* to a maximum value of 50 V rms.

A question now arises concerning the effects of shock durations exceeding the range considered above (i.e., t > 5 s). There is some very limited information that tends to indicate that the threshold of fibrillation may drop only slightly in the range of 5 to 20 or 30 seconds[7]. For times exceeding this period, there is some evidence that asphyxial changes may increasingly assert their influence and lower the threshold further.

There is much less discussion in the literature on the current-time range at which fibrillation ceases and cardiac arrest and respiratory inhibition occurs. Kouwenhoven states that it is in the range of several amperes[2], but mentions no related time period. Because of the lack of substantial information on this aspect, a cautious approach is required. In connection with industrial safety, assuming a 68-kg man, R.H. Lee[3] presents a table in which a generalized value of 4 amperes is given as the threshold of heart paralysis (no fibrillation).

The response of the heart to electrical stimuli is known to vary with different phases of the cardiac cycle. This is a subject that has received considerable attention in basic shock research but has limited significance in safety engineering because in actual shock cases the initiation of exposure with respect to the cardiac cycle is a random variable over which there is no practical control. Also, many shocks may span one or more times the sensitive phase of the cardiac cycles, so in general it must be assumed that the heart will be exposed during its sensitive phase. However, with very short duration shocks, such as result from impulse current produced by capacitor discharges, there is a good probability that they may not occur during a sensitive phase of the cardiac cycle.

A discussion of the relative susceptibility of the heart during various phases of the cardiac cycle is presented in Reference 1. The discussion is based on referenced physiological information and supplemented by shock measurements on animals.

Figure 4-7 shows a photograph of the electrocardiac cycle of a sheep with the various phases indicated in accordance with common medical practice. The following resume of information from Reference 1 is presented for general interest.

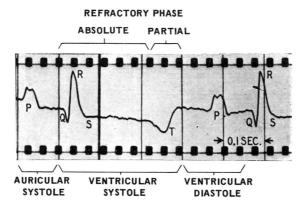

FIGURE 4-7

Phases of Electrocardiac Cycle

The heart is most sensitive to fibrillation from shocks occurring during the partial refractory phase which occurs concurrently with the T wave of the electrocardiogram. This critical phase has a duration of only about 20 percent of the full cycle. Also, the middle of the refractory phase is more sensitive than its beginning or end. Ferris *et al* conclude that fibrillation will not occur with shocks not exceeding about 0.1 second duration unless the shock coincides in part with the T wave of the cardiac cycle.

HIGHER FREQUENCY CURRENTS (kHz RANGE). The effects on the human body of sinusoidal higher-frequency currents and short-duration repetitive impulse currents have received attention principally from the medical profession. Any reader interested in this phase of electric shock should explore medical literature for information.

Although the body appears to be much more tolerant to such exposure, there are unique hazards that have been described by H.B. Williams, M.D. - a member of the L.P. Ferris *et al* investigating team. The following is a quote from Dr. William's remarks presented at the end of Reference 12.

> "On sinusoidal high-frequency alternating currents, or on repeated current pulses of very short duration, account must be taken of the fact, known to physiologists, that as the shock duration decreases, its strength must be increased in order to produce the same stimulation. As the duration becomes very small, this increase must be very great, finally becoming so great that destruction of living substances may occur before it can respond. At higher frequencies, large currents may pass without causing stimulation of muscles or nerves, and these may cause deep heating effects. Since the heat-sensitive mechanism is located in the skin, there is a possibility of damage to internal organs by high-frequency currents, even though no very unpleasant sensations may be apparent. The currents necessary to produce this effect would be in the order of an ampere or more. High-frequency currents of several hundred milliamperes are used quite commonly by the medical profession for deep heating. This form of treatment is called medical diathermy."

4.7 DC CURRENTS

Human exposure to direct current is relatively infrequent compared to low frequency sinusoidal power current. However, there are sufficient areas of possible exposure, such as dc welding equipment, high voltage rectifiers, and carrier repeater power current on communication conductors, to warrant consideration. (Impulse current from capacitor discharges will be covered in a later section.)

Ferris *et al*[1] observed that for shock durations of one second or longer, the level of current causing fibrillation in animals is about five times higher with direct than with alternating current. (All ac values are rms unless otherwise indicated). These investigators further stated that the dc/ac ratio would approach unity for shock durations of a small fraction of a second.

Dalziel[12] theorized that for very short shock times, fibrillating currents will be related to peak values giving a dc/ac ratio of 1.4:1.

The measured data obtained by Knickerbocker[14] concerning direct and alternating current provides significant additional information. The dc/ac fibrillating current ratios for various shock durations are given in Table 4-5.

TABLE 4-5

Fibrillating Current Ratios for Various Shock Durations

TIME DURATION OF SHOCK IN MILLISECONDS	THRESHOLD OF FIBRILLATION		$I_{dc}/I_{ac\,(rms)}$
	AC(RMS) CURRENT* IN MILLIAMPERES	DC CURRENT * IN MILLIAMPERES	
100	900	950	1.06
200	400	500	1.25
500	135	380	2.8
1000	95	350	3.7
2000	80	295	3.7

* These are median values obtained from curves by Knickerbocker[14].

It can be seen that, for a time duration of 0.1 second, the I_{dc}/I_{ac} ratio is practically unity. The Knickerbocker data also shows that in the range from 1 to 2 seconds the ratio remains constant at 3.7. It would be expected that this ratio would not increase significantly during periods of a few more seconds.

A ratio value of 5:1 (based on Ferris' figures) has been frequently used for assessing relative shock hazards. These figures require some reconsideration in view of the Knickerbocker data, which provide more definite values, especially at the shorter durations. It would appear that ratios from 3.5 to 4.0 are perhaps more representative for shock duration times of 1 second and longer. Also, no distinction should be made between dc and ac with regard to shock for duration times of about 0.1 second and less. The same dc/ac ratio may be assumed for dc release currents.

4.8 SURGE CURRENTS OF VERY SHORT DURATION

This subsection presents criteria for judging the possible shock hazard of impulse and oscillatory surges with decay time constants less than about 0.1 second.

It is generally accepted that, in the case of short duration shocks, the energy (expressed in Joules or watt·seconds) to which a victim is exposed is the major factor in assessing shock severity. Other related quantities such as current magnitude, coulombs, and duration do not seen to provide as consistent an index. Dalziel[13] has suggested that the energy concept is suitable for evaluating the hazards of short time exposure to power frequency current as well as to impulse type current.

The statement appears in Bell System literature that, if the energy to which an average human is exposed does not exceed approximately 50 Joules, the resultant shock is unlikely to damage the heart. (The original source of this information is not certain but may have been Ferris *et al.*) Consequently, the following expressions roughly establish the nonlethal shock range for exponential type nonoscillatory surge currents:

$$\text{Energy from a Capacitor (Joules):} \quad 1/2\ CE^2 \leqslant 50, \qquad (2)$$

$$\text{Energy from an Inductor (Joules):} \quad 1/2\ LI^2 \leqslant 50, \qquad (3)$$

where

C = capacitance in farads,

L = inductance in henries,

E = maximum charge potential in volts,

I = maximum current in amperes prior to interruption of the circuit.

Dalziel suggests that the effect of an oscillatory surge current is about twice as severe as in the nonoscillatory case. On this assumption, the tolerable energy level for an oscillatory surge current given by the above expression would be approximately 25 Joules. However, more recent experimental data obtained by Knickerbocker[14] (see Figure 4-11) indicate that there is no practical difference between ac and dc currents for short times of approximately 0.1 second or less.

Dalziel[13] proceeds on the hypothesis, for which he offers sub-stantial proof, that energy is a basic criterion for the numerical evaluation of short duration shocks. He has estimated the energy by an analytical study of the circuits in 13 cases of nonlethal shock involving nonoscillatory current discharges. In all but three of these cases of impulse shock, the energy discharged through the victim's body exceeded 50 Joules. At some of the higher energy levels, the resulting injuries were severe but not fatal. In only one case, in which the energy was estimated to be 5000 Joules, did permanent injury result (see Table 4-6).

Dalziel offers the following expression as a criterion of acceptable risk for a single, nonoscillatory exponential discharge, and states that it 'should give predictions on the side of safety'.

Risk is acceptable when

$$1/2 \ CE^2 \leqslant 0.054 \ (R_b + R_c) \ , \eqno(4)$$

where

C = capacitance in farads,

E = maximum charge potential in volts prior to discharge,

R_b = body resistance in ohms,

R_c = circuit resistance in ohms.

Expressions (4) and (2) are equivalent when $R_b + R_c$ equal 925 ohms.

In Reference 13, Dalziel presents a curve based on expression (4) using an assumed total resistance of 500 ohms, which in his judgement marks the division between the reasonably safe and the dangerous shock areas from the standpoint of fibrillation. This curve, which is reproduced in Figure 4-8, gives values of non-oscillatory exponentially decaying surge current as a function of the circuit time constant. It is based on a 70-kg human at the 0.5 percentile level; that is, only 1/2 percent of a normal population are likely to be more sensitive to shock effects than indicated on the curve.

The dotted curve in Figure 4-8 was derived by the author to indicate the probable safe and dangerous ranges for damped oscillatory surge currents. It was obtained using the following formula from Reference 13 for the maximum allowable energy in 'short, continuous 60-cycle shocks':

$$1/2 \ CE^2 \leqslant 0.027 \ (R_b + R_c). \eqno(5)$$

TABLE 4-8

Summary of Human Accidents Caused By Impulse Currents

LOCATION	TIME CONSTANT (µs)	VOLTAGE (kV)	ALTERNATING CURRENT (A)	QUANTITY (mC)	ENERGY (Joules)	REMARKS
England	0.99	750	1250	1.2	385	Lichtenberg figures behind ears and on chest. Suffered shock.
United States	2.5	60	120	0.3	9	No trace of discharge on body. Headache for 3 days.
Japan	6.0	50	100	0.6	15	Semiconscious and dizzy for short time.
Japan	7.8	960	1600	12.5	5 000	Lost sight one eye. Suffered pain and shock. No trace of discharge on body.
France	8.3	228	456	3.8	429	Lichtenberg figures. Intense muscular reactions and temporary paralysis of hand.
Japan	62.5	80	160	10.0	400	Partially paralyzed for 3 hours.
United States	100	500	1000	100	25 000	Lichtenberg figures. Intense muscular reactions and pain. Deep burn. Paralyzed for 16 hours. Current pathway from abdomen to feet.
Sweden	1 200	25	42	50	520	Unconscious and paralyzed for short period.
Japan	1 200	5	8	10	21	Burn on sole of foot.
Switzerland	3 200	17.5	30	96	720	Unconscious. Wounds on arm and hand.
Switzerland	4 070	17.5	16	66	264	Two men in series, both had wounds.
Sweden	11 180	6	12	134	402	Burns on heels.
Sweden	35 000	2	4	140	140	Concussion due to fall.
Sweden*		1/2 to 1			24	Electrocuted. Small burn on Finger.

* Oscillatory discharge.

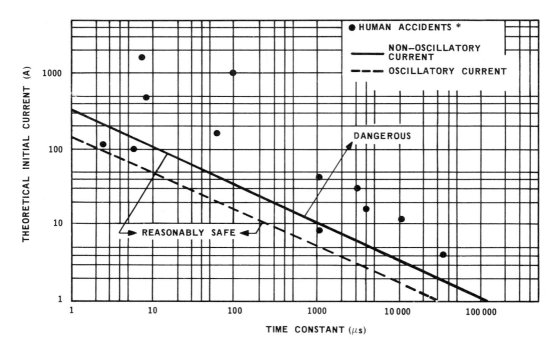

* DOTS INDICATE VALUES OBTAINED FROM STUDY OF NON—FATAL
HUMAN ACCIDENTS [13] INVOLVING NON—OSCILLATORY SURGES.

FIGURE 4-8

Curve for Assessing Electric Shock Hazard
in Cases of Short Duration Surge Currents

Only one fatal shock case was investigated. The body path was from hand to hand, the current waveshape was a damped sinusoid, and the victim received approximately 24 Joules. It was estimated that the circuit voltage was in the range of only 500 to 1000 volts. For this situation, the maximum allowable energy obtained with expression (5) is 13.5 Joules.

Also of interest in Reference 13 are body resistance values obtained in tests conducted by the author. The subjects tested immersed both feet in 3/4 inch of salt water and, with their hands wet with salt water, grasped a 7-gauge copper wire. The minimum resistances observed were 1230 ohms with steady state current and 1135 ohms with impulse current. At the test current levels used, there was no evidence of skin breakdown. Body resistances ranging from 500 to 1500 ohms are frequently used for estimating shock hazards, but the specific value selected by the estimator depends on his personal assessment of probable contact conditions.

In the following example, a rough assessment will be made of shock severity from exposure to capacitor discharges, using the two different approaches that have been discussed.

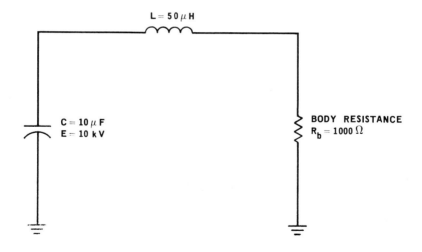

FIGURE 4-9

Capacitor Discharge Circuit

Discharge current is nonoscillatory because

$$r > 2\left(\frac{L}{C}\right)^{\frac{1}{2}} .$$

Method 1:

$$1/2 \; CE^2 = \frac{10^{-5}(10^4)^2}{2} = 500 \text{ Joules (watt} \cdot \text{seconds)}$$

Method 2:

$$\text{Max I at start of discharge} \approx \frac{10\,000}{1000} = 10A$$

$$
\begin{aligned}
\text{Time Constant } T &= C \times R \\
&= 10^{-5} \times 10^3 \\
&= 10^{-2} \text{ s}
\end{aligned}
$$

For reasonable safety (Figure 4-8), the initial current should not exceed about 3.5 amperes.

The conditions of this problem thus constitute a very hazardous shock situation according to both methods of assessment.

Dalziel's curve appears to provide a good basis for assessing the hazard of current surges. His analysis of actual shock cases is interesting because it gives information not generally published and provides a useful, though very approximate, comparison between experience and theoretical concepts.

4.9 ALTERNATING (20 Hz) IN COMBINATION WITH DIRECT CURRENT

4.9.1 General

Minimum fibrillating levels associated with exposure to 20 Hz and direct currents were recently assessed in a study[14] conducted by G.G. Knickerbocker of Johns Hopkins University. (The study was subsidized in part by Bell Telephone Laboratories.) Although the investigation dealt primarily with telephone subscriber ringing circuits, produced typically by a 20 Hz sinusoidal voltage (80-90 V or higher) in series with a nominal 50 volt battery, the results are also applicable to other situations, including the dc offset of sinusoidal power frequency current that occurs during power line fault.

The paper is particularly interesting because it deals with an area of electric shock previously unexplored and because it is one of the more recent papers involving substantial exploratory work with live dogs in the fibrillating current range.

Cumulative distribution diagrams are used to present threshold fibrillating current versus time for 20 Hz and direct currents applied

separately and in combination. However, only the results obtained
with combined currents will be discussed here. The separate ac and
dc data are useful for comparing the severity of ac and dc shock
currents and have been included with the data of other investigators
in Subsection 4.7.

Those interested in the complete details of the test procedure
should refer directly to the paper. However, some of the more
pertinent aspects are outlined below.

1) The dogs varied in weight from 7 to 24 kg, with the majority
   between 10 and 16 kg. Their average weight is not given,
   making weight scaling for data comparison with other
   investigators uncertain. However, as will be seen,
   Knickerbocker summarizes his data in a form that does not
   require the scaling of animal weights up to human levels in
   estimating the shock effect on humans of combined current.

2) Contact was made between the left foreleg and right hindleg.
   This path was intended to correspond to a hand to foot contact in humans, which Knickerbocker states is the most
   common electric shock current path.

3) The shock duration times studied were from about 100 to 200
   milliseconds and from 500 to 2000 milliseconds. The 2000
   millisecond data is of particular interest because it cor-
   responds to the energized phase of the typical ringing
   cycle (2 seconds on, 4 seconds off).

4) There is no data between 200 and 500 milliseconds, but this
   does not significantly detract from the usefulness of the
   results.

5) The start and the duration of the shock were controlled.
   However, no attempt was made to coordinate the start of the
   shock with a preset phase of the ac sinusoid. This random
   application of the ac was probably compensated for by the
   substantial number of tests made.

The statistical procedures employed in analyzing the data are
rather extensive, so the details will not be discussed in this text.
However, some specific aspects requiring comment are as follows:

   No attempt was made in the final analysis to correct for
   variations in animal weight and heart rate. The investigator felt that, in the series of animals studied, the data
   did not show sufficiently strong correlation between these
   variables and fibrillation thresholds to significantly
   affect the results.

4.9.2 Application of Results to Shock Problems

The most important information from the study is presented by Knickerbocker in the following working hypotheses:

1) The hazard associated with a shock consisting of an arbitrary combination of alternating current and direct current, one-half second or longer in duration, is of the same order as that of a pure ac shock of the same peak-to-peak current.

2) The hazard associated with combination shocks of shorter duration, 200ms in particular, is of approximately the same order as that of a pure ac shock of the same peak current. (The data available for 100 ms duration neither confirms nor denies the 'peak hypotheses'.)

The above relationships now provide a practical connection between combination shock currents and the existing literature on human shock which is based largely on pure sinusoidal currents.

The first step in correlating combination shock currents with pure sinusoidal currents is to determine the equivalent peak or peak-to-peak current, as the duration of the combined wave requires, using the following expressions:

$$I_p = I_{dc} + \sqrt{2}\,(I_{ac})$$

$$I_{p-p} = \text{the greater of} \begin{cases} I_{dc} + \sqrt{2}\,(I_{ac}) \\ \text{or} \\ 2\sqrt{2}\ I_{ac} \end{cases}$$

where I_{ac} is an rms value

After obtaining either the peak or peak-to-peak value of equivalent pure alternating current, this value may be converted to an rms value in the conventional manner. This rms value is then in a form that will relate directly to curves and other data in published literature on human shock effects, which for the most part are expressed in terms of pure sinusoidal currents. Although the above relationships were derived from 20 Hz data, they should be applicable to currents in the common power frequency range, since it has already been established by other investigators that there is no significant difference in shock effects between 20 and 60 Hz currents. Also, it is accepted that there is good correlation between the shock behavior of dogs and that of humans. Consequently, the relative reactions between pure and combined currents observed in this series of tests should also apply on a relative basis to humans.

The method of determining a single shock current that is equivalent to a combination of direct and alternating currents is illustrated in Figures 4-10a and 4-10b for representative cases.

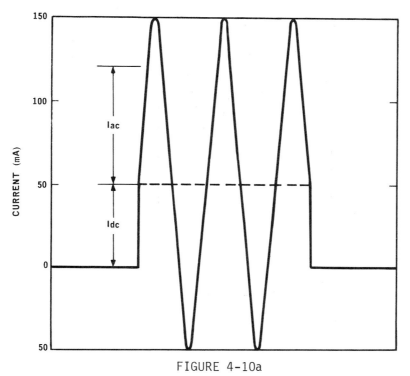

FIGURE 4-10a

Determination of Equivalent Currents - Example 1

Equivalent peak current:
(For durations $\leqslant$ 200 ms)

$$I_p = I_{dc} + \sqrt{2}\ I_{ac}$$

$$= 50\ \ + \sqrt{2}\ (70.7)$$

$$= 150\ \text{mA}$$

$$(I\ \text{rms} = 106\ \text{mA})$$

Equivalent peak-to-peak current:
(For durations $\geqslant$ 500 ms)

$$I_{p-p} = \text{greater of} \begin{cases} I_{dc} + \sqrt{2}\ I_{ac} \\ \quad\text{or} \\ 2\sqrt{2}\ \ I_{ac} \end{cases}$$

$$= 2\sqrt{2}\ \ (70.7) = 200\ \text{mA}$$

$$(I\ \text{rms} = 70.7\ \text{mA})$$

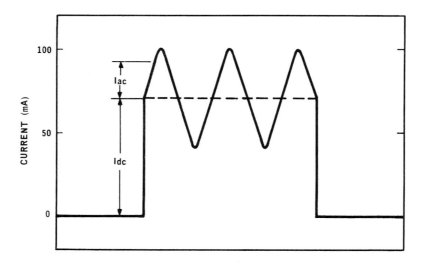

FIGURE 4-10b

Determination of Equivalent Circuits - Example 2

Equivalent peak current:
(For durations ≤ 200 ms)

$$I_p = I_{dc} + \sqrt{2}\, I_{ac}$$

$$= 70 + 30 = 100 \text{ mA}$$

(I rms = 70.7 mA)

Equivalent peak-to-peak current:
(For durations ≥ 500 ms)

$$I_{p-p} = \text{greater of} \begin{cases} I_{dc} + \sqrt{2}\, I_{ac} \\ \text{or} \\ 2\sqrt{2}\, I_{ac} \end{cases}$$

$$= 70 + 30 = 100 \text{ mA}$$

(I rms = 35 mA)

The application of Knickerbocker's data to shock evaluation in humans is demonstrated in the following representative problem involving a combination of direct and alternating currents.

PROBLEM: To determine whether the ringing current typically employed in the Bell System is likely to produce a shock sufficiently severe to cause ventricular fibrillation.

CONDITIONS:

1) DC source = 50 V.

2) AC source = 90 V rms, 20 Hz.

3) Series impedance between source and victim is zero (this is a very conservative assumption, because at most locations in the plant there will be significant series impedance).

4) Contact resistance = 1000 Ω (relatively low value).

5) Body weight = 50 kg (more representative of a female than male).

6) Duration of shock = 2 seconds (based on typical ringing cycle of 2 seconds on and 4 seconds off).

The first step is to determine the alternating and direct currents that will flow through the victim:

$$I_{dc} = \frac{50 \text{ V}}{1000 \text{ }\Omega} = 0.05 \text{ A,}$$

$$I_{ac} = \frac{90 \text{ V}}{1000 \text{ }\Omega} = 0.09 \text{ A.}$$

Since the duration of shock is 2 seconds, an equivalent pure ac current will be determined on a peak-to-peak basis:

$$I_{p-p} = \text{greater of} \begin{cases} I_{dc} + \sqrt{2} \ I_{ac} \\ \quad \text{or} \\ 2\sqrt{2} \ I_{ac} \end{cases}$$

$$= 2.83 \ (0.09) = 0.255 \text{ A}$$

$$I_{rms} = 0.09 \text{ A.}$$

(It is obvious in the case selected that, because the ac is so much greater than the dc component and because dc has much less shock effect, the latter is a minor factor.)

We will now use the Dalziel and Lee formulas (Section 4.6) to determine the range of threshold fibrillation:

$$I_1 = \frac{116}{\sqrt{2}} = 0.082 \text{ A; } I_2 = \frac{185}{\sqrt{2}} = 0.131 \text{ A.}$$

It can now be presumed that the possibility of lethal shock is extremely small. Such a conclusion is supported by the fact that the body current under a very conservative set of assumed conditions is only slightly above the 1/2 percentile minimum fibrillating current line.

This conclusion is also confirmed by experience.

4.10 SUMMARY OF ELECTRIC SHOCK EFFECTS ON HUMANS

The probable ranges of human body currents that produce various degrees of electric shock up to the threshold of fibrillation are summarized in Table 4-7. Values are given for sinusoidal (20 to 60 Hz) and direct currents with respect to a 50 kg (110 lb) body weight. Minimum values are more representative of women, as it seems reasonable that a typical male craftsman, because of his greater body weight, would be able to withstand levels substantially above the minimum given in the table.

Data from Table 4-7 and from R.H. Lee's paper[3] have been plotted in Figure 4-11. Generalized values of this type provide an initial indication of the probable boundaries of shock hazards. Ultimately, however, each problem should be analyzed and assessed on an individual basis using the more complete information provided in the text.

TABLE 4-7

Degree of Shock in 50 kg (110 lb) Human versus ac(20-60 Hz) and dc Currents

DEGREE OF SHOCK	TIME DURATION (SECONDS)	ASSUMED BODY RESISTANCE (OHMS)	TYPE OF CURRENT	CRITICAL CURRENT RANGE FOR 50 kg HUMAN (AMPERES)		ASSUMED BODY RESISTANCE VOLTAGE RANGE (VOLTS)	
				MINIMUM	DEFINITE	MINIMUM	DEFINITE
Perception	0.01 – 5.0	20 k[1]	ac dc[2]	0.0005	0.002	10	40
Let-go	—	1500 [3]	ac	0.009	0.021	14	32
		500	ac dc[2]	0.009	0.021	4.5	10.5
Fibrillation	0.01	1500	ac	1.16	1.85	1740	2780
			dc	1.16	1.85	1740	2780
		500	ac	1.16	1.85	580	925
			dc	1.16	1.85	580	925
	0.1	1500	ac	0.37	0.58	555	870
			dc	0.37	0.58	555	870
		500	ac	0.37	0.58	185	290
			dc	0.37	0.58	185	290
	0.5	1500	ac	0.163	0.260	244	390
			dc	0.326	0.520	488	780
		500	ac	0.163	0.260	81	130
			dc	0.326	0.520	163	260

TABLE 4-7 (CONT'D)

DEGREE OF SHOCK	TIME DURATION (SECONDS)	ASSUMED BODY RESIST. (OHMS)	TYPE OF CURRENT	CRITICAL CURRENT RANGE FOR 50 kg HUMAN (AMPERES)		ASSUMED BODY RESISTANCE VOLTAGE RANGE (VOLTS)	
				MINIMUM	DEFINITE	MINIMUM	DEFINITE
Fibrillation	1.0	1500	ac dc^2	0.116 0.348	0.185 0.555	174 522	277 833
		500	ac dc^2	0.116 0.348	0.185 0.555	58 178	92 277
	5.0 – 20.0	1500	ac dc^2	0.052 0.208	0.082 0.328	78 312	123 492
		500	ac dc^2	0.052 0.208	0.082 0.328	26 104	41 164

1) At the threshold of perception, the currents are so low that skin probably retains its protective properties.

2) Ratios of dc/ac currents producing comparable shock effects:

TIME (s)	dc/ac
less than 0.1	1.0
0.5	2.0
1.0	3.0
more than 5.0	4.0

3) At critical let-go currents and above, the initial voltages and subsequent current will probably disrupt the protective properties of the skin, so body resistance is assumed to be that of the tissue (i.e., 500-1500 ohms).

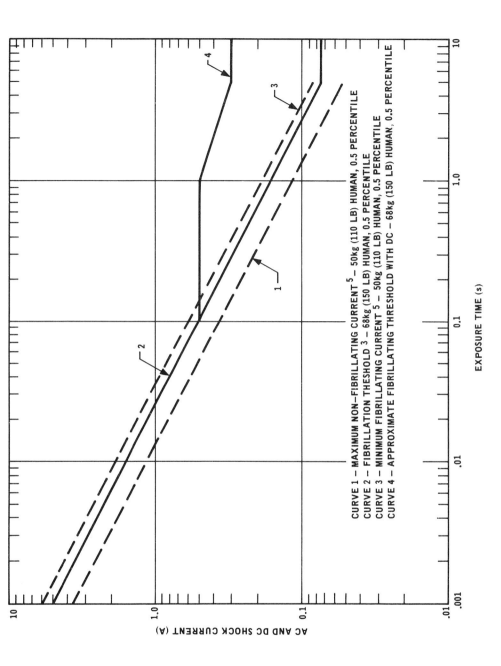

CURVE 1 – MAXIMUM NON–FIBRILLATING CURRENT [5] – 50kg (110 LB) HUMAN, 0.5 PERCENTILE
CURVE 2 – FIBRILLATION THESHOLD [3] – 68kg (150 LB) HUMAN, 0.5 PERCENTILE
CURVE 3 – MINIMUM FIBRILLATING CURRENT 5 – 50kg (110 LB) HUMAN, 0.5 PERCENTILE
CURVE 4 – APPROXIMATE FIBRILLATING THRESHOLD WITH DC – 68kg (150 LB) HUMAN, 0.5 PERCENTILE

EXPOSURE TIME (s)

AC AND DC SHOCK CURRENT (A)

FIGURE 4-11

Generalized Values of Threshold Fibrillating Currents in Humans

4.11 EFFECTS ON HUMANS OF RADIATION FROM MICROWAVE RADIO TRANSMITTERS

R.H. Card (formerly with AT&T, Long Lines Department) presented a paper[16] before the National Safety Congress in 1957 concerning several types of human hazards associated with radio transmitters. His discussion of the effects of microwave radiation on the human body is quoted below.

"In dealing with the safety of high power microwave transmitters, an understanding of the effects of radiation on the human body is helpful in the design of effective measures to minimize this hazard.

The harmful effects are due to internal heating of the body. There is no evidence of ionizing radiation, such as X-rays or gamma rays, from such exposures.

Damage may be done by localized exposure of the eyes and other parts of the body that are not efficiently temperature regulated by blood flow. Possible harmful effects include, among others, the formation of cataracts and damage to the gastro-intestinal and genito-urinary tracts.

Tests have indicated the threshold of field strength producing cataracts in the eyes of rabbits to be in the neighborhood of 1000 watts per square metre for an exposure of about four hours to 12 cm waves.[17] Similar figures for human eyes are not available but it is believed that they are of the same order of magnitude.

Exposure of the entire body to radiation may result in injury or death due to increases in the temperature of the body. Ely and Goldman[18] have experimented with exposure of animals to 10 cm (3000 MHz) waves.[17] Their data indicates that, for dogs, the field strength corresponding to an increase in temperature of one degree C is around 200 to 300 watts per square metre. A further increase of some five or ten milliwatts was found to result in several degrees additional increase in temperature and, in some instances, in death to the animals.

Barron *et al*[19] report the results of examinations of personnel experiencing incidental exposure for varying intervals during the day for a six to nine month period to radiation from radar transmitters. Field strengths ranged from 40 to 130 watts per square metre. Certain changes in the composition of the blood were noted, the significance of which is not known. There were no other effects traceable to these exposures.

Many papers bearing on the subject of radiation effects have been published. The three quoted here are cited as possibly establishing, in order of magnitude, the upper and lower values between which the thresholds of this hazard may lie. Considering only the formation of cataracts from localized single exposures, the magnitude of the threshold is 1000 watts per square metre. But for exposure of the entire body, it appears that the danger level may be a small fraction of this figure."

4.12 ACOUSTIC SHOCK

Acoustic shock occurs when the ear or nervous system is damaged as a result of strong sound pressure on the ear. For example, a telephone or headset user could suffer acoustic shock as a result of a sudden burst of voltage on the telephone receiver. This effect could be produced if a lightning stroke caused nonsimultaneous operation of the protector thereby creating a compensating current through the telephone receiver[20].

The extent of the acoustic shock caused by a disturbing sound or noise depends not only on the loudness of the sound but also on the individual human response to the shock. As is the case with electric shock, the individual response is determined by a number of factors, e.g., physical condition of the victim, time of day, victim's attitude towards the person or device generating the sound, and the characteristics of the sound itself.[21] Consequently, noise ratings are now based on the absolute limit of noise exposures rather than noise levels.

In the new noise control regulations for industry (effective Jan. 31, 1973 under the Canada Labour Code) the absolute limit of noise exposure for zero time exposure is 115 dBA (International Standard A weighting). There is, however, some difference in the loudness perception under free-field or diffused field listening (for which this regulation applied) as opposed to earphone listening. Nevertheless, it appears that the telephone industry can meet the safety requirements if an absolute limit of 115 dB Spl (dB sound pressure level relative to 2×10^{-5} pascal or N/m^2) is set for telephone apparatus.

A device for protection against acoustic shock can be built into the telephone receiver. It consists of two rectifiers of opposite polarities placed in parallel with the telephone receiver. The device eliminates loud bursts of noise while permitting faithful reproduction of the transmitted voice.

Figure 4-12 shows the acoustic output of any 500 type telephone set at an average dc loop current (1000 Hz) versus the signal level available at the line terminals. This figure also shows the voltage burst required to produce a given signal level.

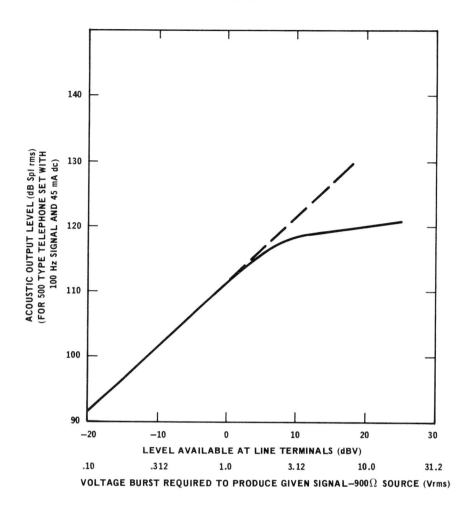

FIGURE 4-12
Acoustic Output of 500 Type Telephone Set

4.13 GLOSSARY

Let-Go Current

The maximum current that an individual can tolerate and still release the energized conductor by using the muscles directly stimulated.

Reaction Current

Current causing involuntary movements.

Threshold of Perception

The level of current flow through some part of the body which initiates an awareness of electric shock. It may produce warmth and a tingling feeling accompanied by some contraction of the muscles.

Ventricular Fibrillation

Disturbed rhythm of the heart with loss of coordination of the organ. In this condition, the circulation of the blood ceases because the heart no longer acts as an effective pump.

4.14 REFERENCES

1. L.P Ferris, B.C. King, P.A. Spence, H.B. Williams, "Effects of Electric Shock on the Heart," *Electrical Engineering,* Vol. 55, May 1936.

2. W.B. Kouwenhoven, "Effects of Electricity on the Human Body," *Electrical Engineering*, March 1949, pp. 199-203. Quoted by permission of Mr. E.K. Gannett, The Institute of Electrical and Electronics Engineers.

3. R.H. Lee, "Electrical Safety in Industrial Plants," *IEEE* Transactions on Industry and General Applications, January/ February 1971.

4. W.B. Kouwenhoven, W.R. Minor, "Field Treatment of Electric Shock Cases," *AIEE* Transactions, Vol. LXXVI, April 1957.

5. C.F. Dalziel, W.R. Lee, "Re-evaluation of Lethal Electric Current," *IEEE* Transactions on Industrial And General Applications, Vol. IGA-4, No. 5, September/October 1968.

6. W.R. Lee, "Death from Electric Shock in 1962-63," *British Medical Journal,* Vol. ii, September 1965.

7. A.P. Kiseleve, "Threshold Values of Safe Current at Mains Frequencies," *Problems of Electrical Equipment, Electrical Supply and Electrical* Measurements (Russion), Sb. MIIT, Vol. 171, 1963.

8. P.H. Gerst, W.H. Fleming, J.R. Malone, "Increased Susceptibility of the Heart to Ventricular Fibrillation During Metabolic Acidosis," Circulation Res., Vol. 19, July 1966.

9. C.F. Dalziel, "Electric Shock Hazards," *IEEE*Spectrum, February 1972.

10. Dalziel, Ogden, Abbott, "Effects of Frequency on Let-Go Currents," *AIEE*, Vol. 62, December 1943.

11. W.R. Lee, "A Clinical Study of Electrical Accidents," *British Journal of Industrial Medicine*, Vol. 18, 1961.

12. C.F. Dalziel, "Dangerous Electric Currents," *AIEE* Transactions, Vol. LXV, August/September 1946.

13. C.F. Dalziel, "A Study of the Hazards of Impulse Currents," *AIEE* Transactions, Vol. LXXII, October 1953.

14. G.G. Knickerbocker, "Fibrillating Parameters of Direct and Alternating (20 Hz) Currents Separately and in Combination - An Experimental Study," *IEEE*, Conference Paper C72-2470, February 1972.

15. H.B. Whitaker, "Electric Shock as Pertains to Electric Fences," *National Fire Prevention Association*, Boston, U.S.A., December 1939.

16. R.H. Card, "The Hazards of Radio Transmitters and Their Correction," *National Safety Congress*, Chicago, Ill., October 24, 1957.

17. D.B. Williams, J.P. Monahan, W.J. Nicholson, J.J. Aldrich, "Biologic Effects Studies on Microwave Radiation: Time and Power Thresholds for the Production of Lens Opacities by 12.3 cm Microwaves," *IRE* Transactions-Medical Electronics, February 1956.

18. T.S. Ely and D.E. Goldman, "Heat Exchange Characteristics of Animals Exposed to 10 cm Microwaves," *IRE* Transactions - Medical Electronics, February 1956.

19. Charles I. Barron, M.D., Andrew A. Love, M.D., Albert A. Baraff, M.D. "Physical Evaluation of Personnel Exposed to Microwave Emanations," *The Journal of Aviation Medicine*, Vol. 26, December 1955.

5. EARTH POTENTIAL GRADIENTS

5.1 INTRODUCTION

When a substance conducts electric current, a potential difference occurs. The **gradient** of such a potential is the rate of increase or decrease of its magnitude with distance from some point of
reference.

The scope of this section is limited to the consideration of
potential gradients appearing in the earth in the vicinity of grounding electrodes. Earth potentials may create a shock hazard, damage
communication plant and apparatus, and interrupt service because of
permanent grounding of carbon block protectors. Major sources of
hazardous earth potentials are lightning discharges to earth or to
buried conducting objects, and earth return power fault currents.

Appendix A to this section describes analytical procedures for
determining potential gradients along the surface of the earth.
Appendix B contains some general information on earth potentials
caused by magnetic storms.

The major situations of specific concern to communication
engineers are:

1) Power ground rods at the foot of joint-use poles. The
    gradient around a rod is sufficiently steep that a person
standing only 18 inches away and touching the grounding conductor to the rod could be exposed to a major portion of the
total earth potential.

2) Power station ground mats. Potential gradients that
    develop around such grounding structures may typically
distribute over distances of a few hundred feet to several
thousand feet. Although the total potential rise distributes
over a substantial distance, potential differences between
two points of 'human reach' or 'step distance' appearing in
the immediate vicinity of the mat may present a shock
hazard. Communication line and cable facilities entering such
areas and extending to terminal points remote from the
station are exposed to the total potential rise of the mat
unless mitigated by protection measures.

Central offices and subscriber stations in the proximity
of a power station mat, though not directly connected to it,
may be exposed to sufficient potential during power line
faults to operate and permanently ground carbon protector
blocks.

3) Potential gradients can appear around the footings of power
   transmission line towers,but, because the physical size of
   the grounding arrangement is generally small compared to a
   station mat, the area affected will be relatively restricted.
   However, the potential developed by a lightning stroke to a
   tower can ionize a discharge path through several feet of soil
   to a buried cable. Also, a small degree of hazard can be
   associated with splicing operations on buried cable conducted
   close to a tower footing because of a local potential rise that
   could result from a line to ground fault.

The potential distribution around a grounding electrode can be
conveniently presented in the form of a curve, normalized so that
the gradient is expressed as a percent of the total voltages appearing at various distances from the edge of the electrode. The shape
of such a distribution curve is exclusively a function of the
geometry of the grounding structure. Absolute magnitudes may vary
with other factors such as soil resistivity and fault current amplitudes, but the percentage change in potential with distance remains
the same.

5.2 METHODS OF DETERMINING EARTH POTENTIAL
     DISTRIBUTION PATTERNS

5.2.1 General

Grounding structures are constructed in a variety of sizes and
configurations, and it is usually difficult to accurately predict
field patterns by analytical methods. This is especially true where
several foreign objects are present that perturb the general symmetry of the situation. In such cases, field measurements (which
are usually made after construction of the electrode) may be needed
to evaluate analytical predictions. Scale model tests are sometimes made using an 'electrolytic tank', which is a vessel containing a liquid compounded to simulate the conductivity of soil. (Field
measurements and scale model tests are discussed briefly later.)

The first step in solving an earth potential problem is to
mathematically analyze a model of the system, typically a hemispherical electrode buried flush with the surface of homogeneous
soil. The principal difficulty with this procedure is in determining the dimension of the hemispherical electrode that will simulate
the potential pattern of the original structure under consideration.
Some information of an empirical nature is available to assist in
the selection of equivalent hemispherical electrodes. Such solutions may at times be subject to substantial error, but in many
cases are adequate for engineering purposes.

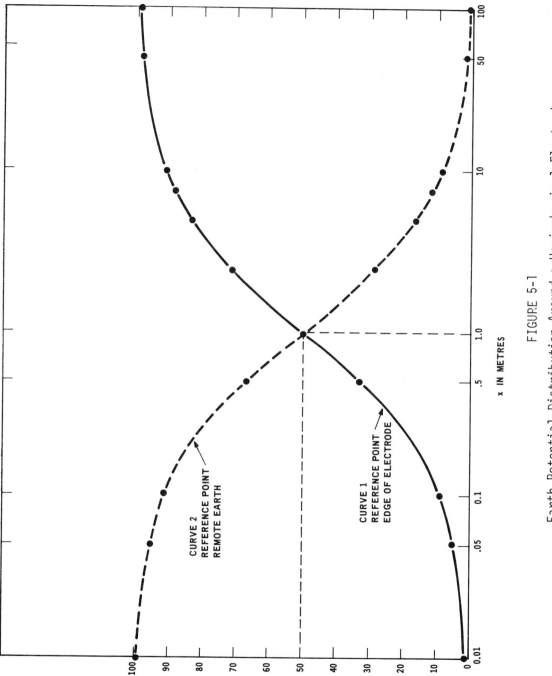

FIGURE 5-1

Earth Potential Distribution Around a Hemispherical Electrode
Having a Radius of One Metre

Simple methods of approximating earth potential magnitudes and their field distribution on or near the surface of the earth may be very useful to an engineer when he is confronted with situations involving changes in the existing relationship between communication plant and power system grounds, or when he is required to make right of way decisions with regard to new plant in the vicinity of power installations.

5.2.2 Basic Analytical Concepts

Expressions for computing the voltage field distribution in the earth around a hemispherical electrode are given here, and the typical shapes of the gradient for different reference points are shown in Figure 5-1.

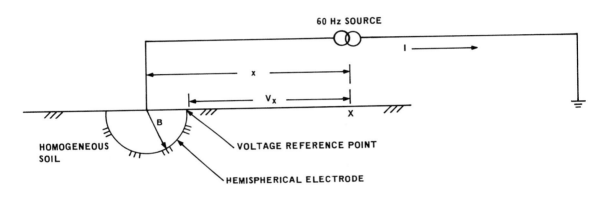

FIGURE 5-2

Determination of Voltage at a Point X
from a Hemispherical Electrode

The voltage at any point x with respect to edge of the electrode is given by[14]

$$V_x = \frac{I\rho}{2\pi} \left(\frac{1}{B} - \frac{1}{x} \right) \tag{1}$$

where

 I = Earth return current to electrode in amperes

 ρ = Earth resistivity in ohms-metre

 B = Radius of hemisphere in metres

 x = Distance in meters from center of hemisphere to point x
     under consideration

The total voltage between the edge of the electrode and a remote point
(x = ∞) is

$$V_\infty = \frac{I\rho}{2\pi B} \quad .$$

(2)

By combining expressions (1) and (2),

$$\frac{V_x}{V_\infty} = 1 - \frac{B}{x} \quad .$$

(3)

The voltage ratio plotted in curve 1 of Figure 5-1 shows how
the potential in the earth distributes when measured from the edge
of the electrode as shown in Figure 5-3.

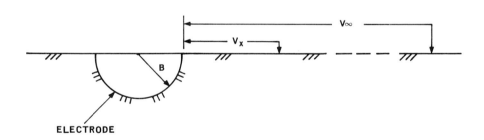

FIGURE 5-3

Determination of Voltage Field Distribution

Presentation of data in the form of curve 1 in Figure 5-1 is useful for solving problems concerning the hazards of 'step and touch potentials'. 'Step potential' is the voltage that may appear between the feet of a person standing on the ground with his feet separated by a normal step length. 'Touch potential' is the voltage occurring between a person's feet and one hand. If we consider step potential, it is evident from curve 1 that exposure increases as one proceeds towards the electrode. A person having a step distance of 0.75 meter positioned between 0.1 and 0.85 meter from the electrode would intercept about 37 percent of the total potential field. However, at a distance of 10 meters away from the electrode, the step potential would not exceed one percent of the total field. As mentioned in Section 4 (Electric Shock) of this Report, the human body is much more severely exposed to shock effects when the contact is between the hands and feet than when the contact is between the feet only, since the former body path more directly involves the heart. This is sometimes not recognized in discussions of touch and step distances. The subject of step and touch potentials in and around power and at tower footings is discussed further in Reference 1.

Perhaps of more interest to the communications engineer is the distribution pattern of voltage as it increases with distance from a remote earth to the elevated potential of a grounding electrode as shown in curve 2 of Figure 5-1. The reference of measurement in this case is a remote ground, as indicated in Figure 5-4.

NOTE: In the case of the hemispherical electrode:

$$V_2 = V_\infty - V_x = \frac{I\rho}{2\pi x} \qquad (4a)$$

and $\quad \dfrac{V_2}{V_\infty} = \dfrac{B}{x}$. $\qquad (4b)$

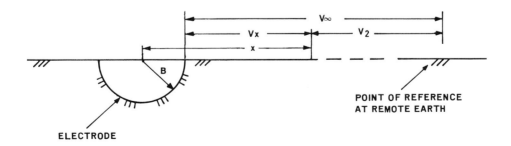

FIGURE 5-4

Voltage Between a Grounding Electrode and a Remote Earth

5.3 FIELD MEASUREMENTS

Bell System operating companies have conducted a limited number of tests with 60 Hz earth return currents in connection with the engineering of neutralizing transformer installations. In these tests, the principal objective was to locate a suitable remote ground for the transformer. Some data were collected that provided an empirical basis for the selection of equivalent hemispherical electrodes to simulate ground mats of various size with regard to the earth potential gradients. This information was incorporated in an AIEE paper[2] presented in 1962 on the general subject of earth potential distributions external to power station grounding structures. The curves in Bell System Practice Section 837-310-100 on the distribution of earth potential from the edge of power ground mats were derived from this source.

To check proper equipment functioning and safety around stations, power authorities occasionally undertake field measurements of existing stations (see References 5, 6, and 8). It is important to note that while actual distributions in the ground deviate somewhat from the calculations because of unknown underground structures, the maximum ground potential rise (GPR) coincides with the predicted value.

5.4 SCALE MODEL TESTS

Some investigations of earth potentials in and around grounding electrodes have been conducted with scale models. The method consists of placing the model under study in an electrolyte large enough to minimize boundary effects. Scaling is done on the size of the model and the resistivity of the solution. (See References 4, 5, and 9)

Scaling ratios of more than 20 introduce physical difficulties and unacceptable measuring errors. Consequently, with large structures having mat areas of several hundred thousand square feet, studies are limited to sections within the confines of the structure or its immediate vicinity. In the study of extended fields outside the mat area, the investigator can only model relatively small structures.

5.5 EXISTING DATA

Because of the cost and the magnitude of the problems involved, existing information on actual measurements or design values are rather scarce. The data provided in Figures 5-5, 5-6, and 5-7 and Table 5-1 represents, to the writer's knowledge, the most pertinent information that is available for engineering purposes.

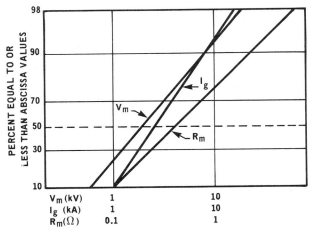

FIGURE 5-5

Distribution of Maximum Voltage, Station Resistance and
Ground Return Current (Survey Results from 2605 Power Stations)[7,10]

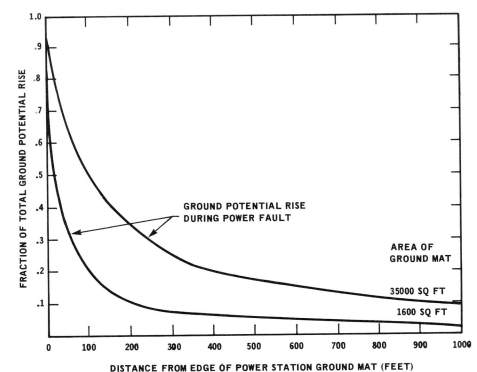

FIGURE 5-6

Typical Distribution of GPR from Edge of Station
According to Area for Small Ground Mats[11]

259

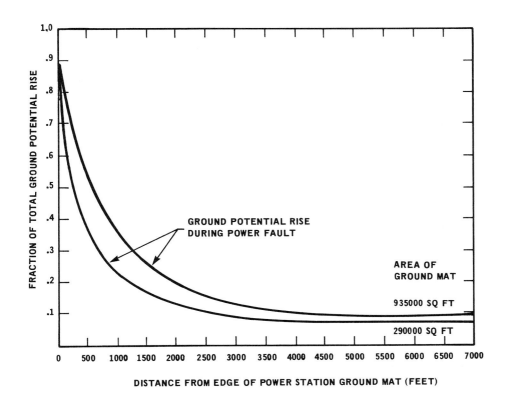

FIGURE 5-7

Typical Distribution of GPR From Edge of Station
According to Area (for Large Ground Mats)
(After Reference 11)

TABLE 5-1

Design Values in Ontario Hydro and Hydro Quebec[13]

DESIGN PARAMETER	ONTARIO HYDRO	HYDRO QUEBEC
GPR at Stations: Typical Maximum	3 kV 5 kV	– 5* kV
Tower Footing Resistances (Including Skywires and Counterpoises)	–	10 Ω

*Still under study

5.6 REFERENCES

1. A. Elek, "Hazards of Electric Shock at Stations During Faults, and Methods of Reduction", *Ontario Hydro Research News*, Jan./ March 1958, Vol. 10.

2. D.W. Bodle, "Earth Potential Distributions Associated with Power Grounding Structures", *AIEE* Paper No. CP 62-205, Winter General Meeting, January 1962.

3. E.D. Sunde, "Earth Conduction Effects in Transmission Systems", *Dover* Publications, N.Y.

4. H.R. Armstrong, "Grounding Electrode Characteristics from Model Tests", *AIEE* Technical Paper 53-398, Sept. 1953.

5. *IEEE* Standard No. 80, "Guide for Safety in AC Substation Grounding".

6. McKenna and O'Sullivan, "Induced Voltages in Coaxial Cables and Telephone Lines", CIGRE Conference Paper No. 36-01, 1970 (Induction).

7. *IEEE* Committee Report, "A Guide for the Protection of Wire Line Communication Facilities Serving Electric Power Stations", PAS 85 No. 10, October 1966.

8. Borgvall and Al, "Voltages in Substation Control Cables During Switching Operations", CIGRE Conference Paper No. 36-05, 1970.

9. Pettersson, "The Influence of Power Lines on Telecommunication Circuits", Ericsson Rev. Tele, 2/1971.

10. B.S.P. 876-310-100, Bell System.

11. B.S.P. 876-310-100CA, Bell Canada.

12. Hydro Quebec Report, "Mésure de la Resistance Pylone des Terres Circuits 1313 and 1322", October 1968. (Available in Dept. 3E30 File No. VII.)

13. Endrenyi, "Fault Current Analysis for Station Grounding Design", Ontario Hydro Research Quarterly, Second Quarter, 1967.

14. Rüdenberg, "Grounding Principles and Practice", Elect. Eng., Vol. 64, No. 1, 1945.

5.7 APPENDIX A: POTENTIAL GRADIENTS ALONG THE SURFACE OF THE EARTH — ANALYTICAL CONSIDERATIONS AND PROCEDURES

A.1 GENERAL

Grounding electrodes may in practice have a variety of configurations but, for the purposes of engineering approximation, they can be classified into three general types, namely, vertical rods, horizontal wires, and mat structures composed of a multiplicity of rods interconnected by a mesh of horizontal wires.

The following assumptions must be made to simplify generalized predictions regarding the behavior of buried electrodes.

1) The soil is homogeneous and therefore has a uniform resistivity throughout the area under consideration.

2) There are no buried conducting objects in the immediate vicinity of the electrodes that can perturb the symmetry of the field.

3) Simple electrode configurations, such as a single rod or a horizontal counterpoise wire, can be considered directly, but for more complicated arrangements, such as power transmission tower footings and mat structures, it is common practice to reduce them to equivalent conducting hemispheres that present a uniform radial field and thus facilitate analytical manipulation.

Most grounding structures used in the power and communication industries are buried at relatively shallow depths. It is permissible, therefore, to assume for purposes of approximation that the electrode is essentially at the surface of the earth. The resistance to earth of a deep electrode can approach one half that of the same electrode at the surface because in the latter case only one half of the electrode has effective contact with the soil. The magnitude of the potential gradient produced on the surface of the earth by a shallow electrode will be substantially greater than if it was placed well below the surface because, for a given current and resistivity, the maximum voltage rise of the electrode with respect to remote earth is greater. Therefore, it is likely that the estimates of earth potential gradients obtained using these usual analytical simplifications will be somewhat higher than those encountered in practice.

From the standpoint of hazard to operating personnel, this leads to conservative predictions. Since many variables are involved in an overall appraisal of shock hazards, the approximation of electrode ground resistance and earth potential gradients in the manner illustrated in the following examples should be sufficiently accurate for most engineering purposes. Where the acceptable accuracy of this approach is questioned because of perturbing factors unique to a specific situation, actual field measurements may be the only alternative.

A.2 THE EARTH POTENTIAL ON THE SURFACE OF THE EARTH AROUND A VERTICAL GROUND ROD[3]*

Method 1

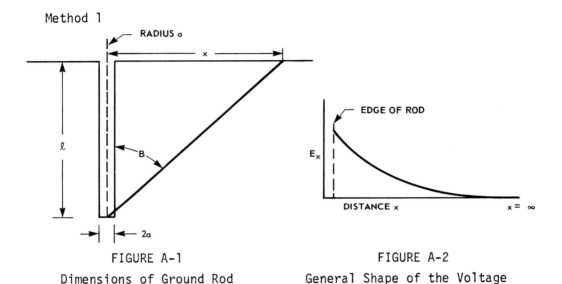

FIGURE A-1

Dimensions of Ground Rod

FIGURE A-2

General Shape of the Voltage Gradient Around the Rod

Sunde[3] (page 90, section 3.13) gives the following general formula for computing the approximate voltage with respect to a vertical rod electrode that appears on the surface of the earth when the electrode is energized:

$$E_x = \frac{I\rho}{2\pi\ell} \ \log_e \frac{(\ell^2 + x^2)^{\frac{1}{2}} + \ell}{x} \qquad (5)$$

where

I = current in amperes

ρ = soil resistivity in ohms-metre

ℓ = length of rod in metres

x = distance from center of rod in metres

* References for Appendix A are listed with those for the main part of text. (See subsection 5.6)

Now the total potential of the rod, E_r, is given by the following expression:

$$E_r = \frac{I\rho}{2\pi\ell} \log_e \frac{2\ell}{x} \quad \text{when } a << \ell. \tag{6}$$

Where a is the radius of the rod, and consequently x is much smaller than its length, ℓ. Ground rods very often have dimensions ranging in diameter from 1.27×10^{-2} to 2.54×10^{-2} metres and lengths from 2.44 to 3.05 metres. With electrodes of such dimensions, the earth potential drops off rapidly with distance away from the rod, which will be illustrated in the following example. Using the above formulas, the percent of total voltage that appears at a distance from a rod equivalent to its length will be computed for the following representative case.

Conditions: I = 1 ampere

ρ = 100 ohms-metre

ℓ = 2.44 metres

a = 0.955×10^{-2} metres

x = ℓ = 2.44 metres

Therefore,

$$E_x = \frac{1 \times 100}{6.28 \times 2.44} \log_e \frac{(2.44^2 + 2.44^2)^{\frac{1}{2}} + 2.44}{2.44}$$

$$= \frac{100}{15.32} \log_e \frac{3.45 + 2.44}{2.44}$$

$$= 6.527 \log_e 2.414$$

$$= 6.527 (.88) = 5.74 \text{ volts}$$

$$E_r = \frac{1 \times 100}{6.28 \times 2.44} \log_e \frac{2 \times 2.44}{.955 \times 10^{-2}}$$

$$= 6.527 \log_e 511 = 6.527(6.235) = 40.7 \text{ volts}$$

Now,
$$\frac{E_x}{E_r} = \frac{5.74}{40.7} = 0.141 \quad \text{or} \quad 14\%.$$

The same results may be obtained with the methods outlined by Rüdenberg[14]:

$$E_x = \frac{I\rho}{2\pi\ell} \log_e \cot \frac{B}{2}. \tag{7}$$

Formula (7) is a general expression, good for all distances. When x is small compared to ℓ, then $\cot \frac{B}{2}$ simplifies to $\frac{2\ell}{a}$.

Now,

$$E_x = \frac{I\rho}{2\pi\ell} \log_e \frac{2\ell}{a}, \qquad \text{where} \quad x \ll \ell. \tag{8}$$

Formula (7) produces the same results as formula (5), which may be noted from the following substitution:

$$E_x = \frac{1 \times 100}{6.28 \times 2.44} \log_e \cot \frac{B}{2};$$

$$\cot B = \frac{2.44}{2.44} = 1, \quad B = 45°;$$

$$\cot \frac{B}{2} = \cot 22.5° = 2.414;$$

$$E_x = 6.527 \log 2.414 = 5.74 \text{ volts.}$$

The values in table A-1 were obtained with formula (5) but it has been demonstrated above that identical results could have been obtained with formula (7).

TABLE A-1

Distribution of Earth Potential Around a Vertical
Ground Rod (Radius = 0.955 cm., Length = 2.44 m.)

DISTANCE x (METERS)	PERCENT OF TOTAL VOLTAGE*
0.955×10^{-2}	100.0
0.01	99.0
0.05	73.5
0.10	62.4
0.25	36.6
0.50	28.7
1.00	26.0
2.00	15.2
2.44	14.1
3.00	13.3
4.00	9.2

* With respect to 'remote' earth.

The curve in Figure A-3 illustrates that the earth potential drops very rapidly close to the rod, which is significant from a shock standpoint, e.g., if a person standing 1 meter away touches the rod or a conductor to it, this person will be exposed to about 74% of the total potential rise of the rod. At greater distances the voltage drops off rather slowly and some small percentage remains even at a substantial distance from the rod. In multi-rod arrangements this is the reason for the mutual effects that reduce paralleling efficiency.

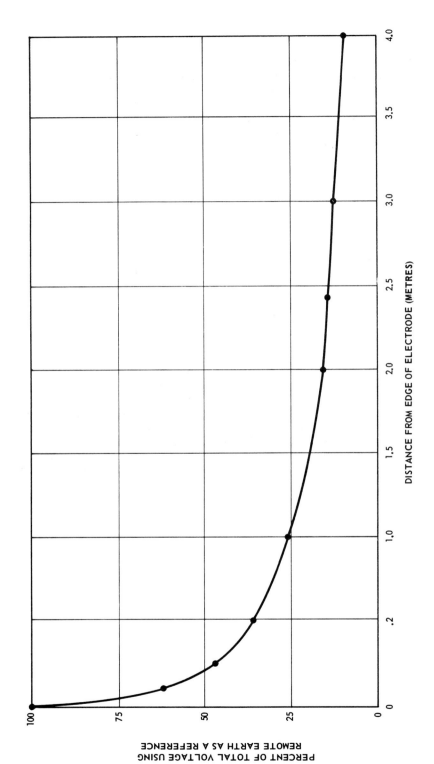

FIGURE A-3

Decrement of Potential Field Along Surface of the Earth for a
Vertical Ground Rod (Rod Diameter = 0.955 cm, Length = 2.44 m)

A.3 POTENTIAL GRADIENT AROUND A SMALL MULTIPLE ROD GROUNDING ARRAY[2]

Method 1 - Equivalent Hemisphere on Basis of Comparable Resistance

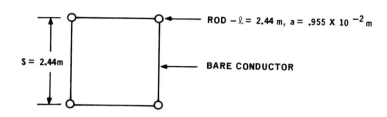

FIGURE A-4a

Multiple Rod Grounding Array

Radius, A, of a single equivalent rod (see Reference 3)

$$A = (a\sqrt{2}\ S^3)^{\frac{1}{4}}$$

$$= [0.995 \times 10^{-2}(1.41)2.44^3]^{\frac{1}{4}}$$

$$= (0.195)^{\frac{1}{4}} \qquad\qquad (9)$$

$$= 0.665\ m$$

where

a = radius of rod in metres

S = distance between two rods in metres

The potential gradient around this equivalent rod was obtained with equation (5) and is plotted in Table A-2. Equivalent hemisphere methods such as those illustrated in the following examples provide a simple way of approximating potential gradients around grounding arrays.

Method A - Equivalent Hemisphere on Basis of Comparable Resistance

Now, resistance of the rod,

$$R_A = \frac{\rho}{2\pi\ell} \log_e \frac{2\ell}{A} \qquad (10)$$

$$= \frac{100}{6.28 \times 2.44} \log_e \frac{4.88}{0.665}$$

$$= 13\Omega.$$

Radius of equivalent hemisphere,

$$B = \frac{\rho}{2\pi R_A}$$

$$= \frac{100}{(6.28)(13)}$$

$$= 1.22 \text{ metres.}$$

The field distribution around a hemisphere of $B = 1.22$ metres using equation (4b) is given in Table A-3.

Method B - Equivalent Hemisphere on the Basis of Comparable Volumes

This method consists of determining the radius, B, of a hemisphere having the same volume as the soil enclosed within the grounding array. In this example, the volume is 2.44^3 or 14.5 cubic metres.

Radius of the hemisphere,

$$B = (\frac{3Vg}{2\pi})^{\frac{1}{3}} , \qquad (11)$$

where

Vg = enclosed volume of grounding structure in cubic metres.

In this example,

$$B = [\frac{3(14.5)}{6.28}]^{\frac{1}{3}}$$

$$= 1.91 \text{ metres.}$$

The values given in Table A-4 were also obtained using equation (4b).

In practice, possible unknown variables make the accuracy of any format computation somewhat uncertain. Consequently, any of these methods are suitable for initial approximations.

A.4 POTENTIAL GRADIENTS AROUND LARGE GROUNDING STRUCTURES

Empirical methods of approximating the distributions of earth potentials around large grounding structures of the type used at power stations (Figures A-9 and A-10) are discussed in Reference 2. In this section three methods of equating the dimensions of typical grounding structures (lattice mats supplemented with vertical rods) to equivalent hemispheres are tested by comparing analytical results with measured values.

> *Note:* Reference 5 contains a list of papers dealing with the subject of grids, mats, and plates resistances as well as potential distributions. The data in Tables A-2, A-3, and A-4 are plotted in Figure A-4b to facilitate comparison.

TABLE A-2

Potential Gradient Around Equivalent Rod
(Equivalent Radius = 0.665 m, Length = 2.44 m)

DISTANCE FROM EDGE OF ROD (m)	x (m)	PERCENT OF TOTAL VOLTAGE
0.00	.665	100.00
0.05	.715	96.5
0.10	.765	93.5
0.50	1.165	74.1
1.00	1.665	58.5
2.00	2.665	40.8
3.00	3.665	31.2
5.00	5.665	20.8
7.00	7.665	15.7
10.00	10.665	11.1

TABLE A-3

Decrement of Voltage for a Multiple Rod
Grounding Array (Method A)

$\ell = (X-B)$ (m)	X (m)	PERCENTAGE OF TOTAL VOLTAGE*
0.00	1.22	100.0
0.05	1.27	96.0
0.10	1.32	92.5
0.50	1.72	71.0
1.00	2.22	55.0
2.00	3.22	38.0
3.00	4.22	29.0
5.00	6.22	19.6
7.00	8.22	14.8
10.00	11.22	10.9

* With respect to remote earth.

TABLE A-4

Decrement of Voltage for a Multiple Rod
Grounding Array (Method B)

$\ell = (X-B)$ (m)	X (m)	PERCENTAGE OF TOTAL VOLTAGE*
0.00	1.91	100.0
0.05	1.96	97.5
0.10	2.01	95.0
0.50	2.41	79.0
1.00	2.91	66.0
2.00	3.91	49.0
3.00	4.91	39.0
5.00	6.91	27.6
10.00	11.91	16.0

* With respect to remote earth.

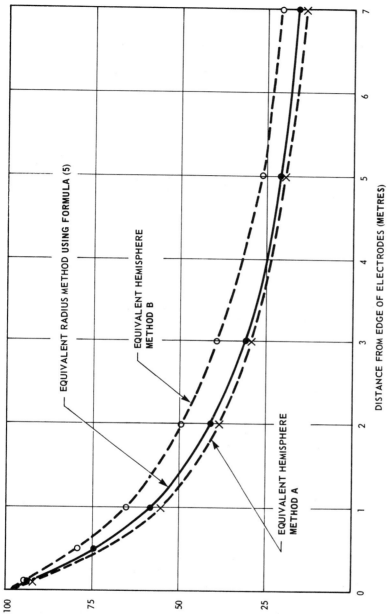

FIGURE A-4b

Decrement of Potential Field Along Surface of the Earth
for the Multi-rod Arrangement in Fig. A-4a

Method 1

Compute the volume, V_G, of the grounding structure by multiplying the mat area by the length of vertical rods. Then find the radius, B, of the hemisphere of equivalent volume from expression (11).

Method 2

Approximate the effective area of the grounding structure A_G by assuming it to be the area of the mat plus the area of the sides to the depth of the ground rods. The radius, B, of a hemisphere having an equivalent surface area is

$$B = (\frac{A'_G}{\pi})^{\frac{1}{2}} . \tag{12}$$

Method 3

Determine the area A'_G of only the horizontal ground mat. Then find the radius of the hemisphere from equation (12):

$$B = (\frac{A'_G}{\pi})^{\frac{1}{2}} .$$

The radius of this hemisphere is the same as the radius of a circle having an area equal to that of the horizontal ground mat.

With the exception of very small mat areas, Methods 2 and 3 appear to provide a better agreement of measured and computed values.

Method 1 provides values that most closely check the measured values for a 1600 sq ft mat area (see Figure A-5). Method 2 provides results that are somewhat more conservative but still satisfactory for most purposes.

In the case of the 35 000 sq ft mat, the curve obtained with Method 3 practically coincides with measured values (see Figure A-6). Method 2 also provides acceptable results.

The results obtained by Methods 2 and 3 are both acceptable for a structure of 290 000 sq. ft. area (see Figure A-7).

The measured data for the 935 000 sq ft structure (Figure A-8) are very limited, consisting of only four points relatively remote from the station mat. The scope of the test was confined to investigating remote locations suitable for grounding the primary of a three winding neutralizing transformer. In this situation, it is likely that computations employing either Method 2 or 3 would have been useful in developing an analytical solution.

The data presented in Table A-2 are for short distances from the station fence, which was bonded to the ground mat. Measurements were made at two locations around the station perimeter using a 5 ft ground rod for the electrode. Open circuit voltages were measured initially and then repeated with a 1500 ohm shunt across the high impedance voltmeter. These latter data indicate that the coupling presents a fairly low impedance path, since the placing of a 1500 ohm shunt across the measuring circuit only reduced the open circuit voltage by about 20 percent.

It may be noted that in all of the comparisons presented there was a general tendency of the computed method to indicate a some-what more gradual drop in potential with respect to remote earth on moving away from a grounding structure than was observed by measurements.

A.5 POTENTIAL DISTRIBUTION IN THE VICINITY OF A BURIED HORIZONTAL CONDUCTOR

The distribution of potential on the surface of the earth in the vicinity of a long horizontal conductor such as a buried counterpoise can be obtained with the following expression obtained from Reference 3, subsection 5.5. (See also Figure A-12.)

Case No. 1 Voltage normal to the conductor along y axis with x = 0,

$$E_{(0,y)} = \frac{I\rho}{\pi\ell}\log_e \frac{[(\frac{\ell}{2})^2 + y^2 + d^2]^{\frac{1}{2}} + \frac{\ell}{2}}{(y^2 + d^2)^{\frac{1}{2}}} \ . \tag{13}$$

Data for a typical situation is given in Table A-5 and plotted in Figure A-13. It should be noted that expression (13) gives voltage with respect to a remote measuring point; consequently, voltage decreases as the measuring probe is moved away from point 0.

TABLE A-5

y	VOLTAGE E_y	PERCENT VOLTAGE WITH RESPECT TO REMOTE EARTH
0.00	0.376	100.0
0.10	0.375	99.8
0.25	0.368	98.0
0.50	0.350	93.0
1.00	0.329	87.5
5.00	0.250	66.7
10.00	0.214	56.8
25.00	0.167	44.3
50.00	0.130	34.7
100.00	0.096	25.5
300.00	0.046	12.4
500.00	0.030	8.0

Assumed Conditions:

$$I = 1 \text{ ampere}$$

$$\rho = 100 \ \Omega \cdot m$$

$$\frac{\ell}{2} = 305 \ m$$

$$d = 0.457 \ m$$

The maximum voltage will appear at point 0 and, when $d \ll \ell$ (as in this case), it may be obtained with the following expression:

$$E_{(0,0)} \simeq \frac{I\rho}{\pi\ell}\log_e \frac{\ell}{d} \ . \tag{14}$$

The data may now be normalized on the basis of percent of total voltage using the ratio $E_{(0,y)}/E_{(0,0)}$, and plotted as in Figure A-13. The general distribution now holds for any soil resistivity with a conductor having the same physical dimensions and configuration.

TABLE A-6

x	E_x	PERCENT OF TOTAL VOLTAGE WITH RESPECT TO REMOTE EARTH
0.00	0.376	100.0
100	0.370	98.6
200	0.358	95.0
250	0.346	92.0
275	0.328	87.0
300	0.283	75.0
305	0.206	54.7
310	0.126	33.4
325	0.090	24.0
350	0.070	18.6
400	0.052	14.0
500	0.037	10.0

Assumed Conditions:

$$I = 1 \text{ ampere}$$

$$\rho = 100 \ \Omega \cdot m$$

$$\frac{\ell}{2} = 305 \text{ m}$$

$$d = 0.457 \text{ m}$$

Case No. 2 Voltage along the x axis with y = 0

$$E_{(x,0)} = \frac{I\rho}{2\pi\ell} \log_e \frac{[(x+\frac{\ell}{2})^2 + d^2]^{\frac{1}{2}} + (x + \frac{\ell}{2})}{[(x-\frac{\ell}{2})^2 + d^2]^{\frac{1}{2}} + (x - \frac{\ell}{2})} \cdot \quad (15)$$

Voltage distribution along the x axis is shown in Figure A-14 for the same conditions assumed in Case No. 1 (see Table A-6 above).

Case No. 3 Voltage at any point x,y is given by the expression

$$E_{(x,y)} = \frac{I\rho}{2\pi\ell} \log_e \frac{[(x+\frac{\ell}{2})^2 + y^2 + d^2]^{\frac{1}{2}} + (x + \frac{\ell}{2})}{[(x-\frac{\ell}{2})^2 + y^2 + d^2]^{\frac{1}{2}} + (x - \frac{\ell}{2})} \cdot \quad (16)$$

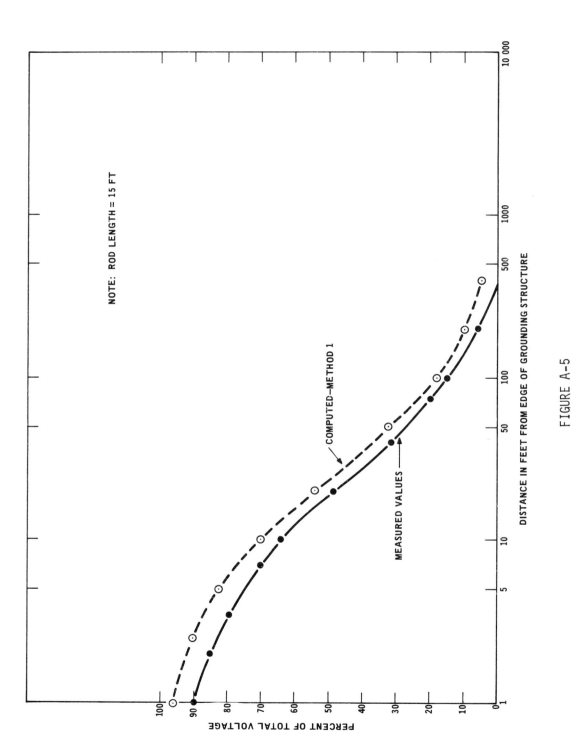

FIGURE A-5

Earth Potential Distribution from the Edge of a Grounding Structure
With Respect to a Remote Earth (Mat Area 1600 Square Feet)

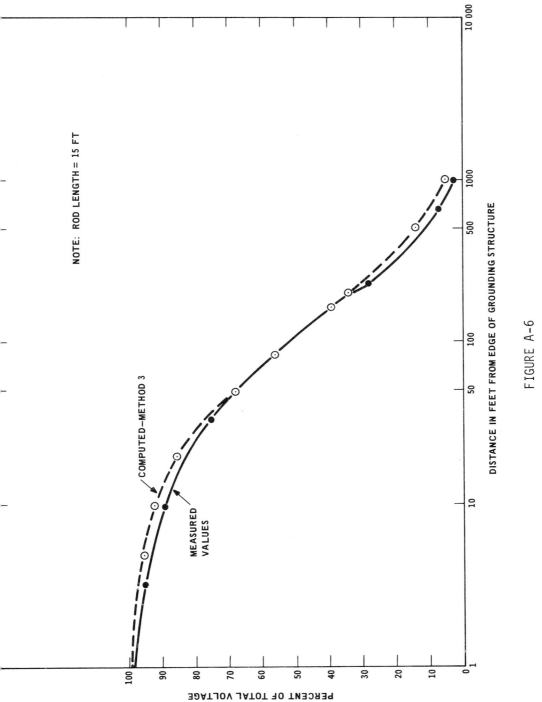

FIGURE A-6

Earth Potential Distribution from the Edge of a Grounding Structure
With Respect to a Remote Earth (Mat Area ≈ 35 000 Square Feet)

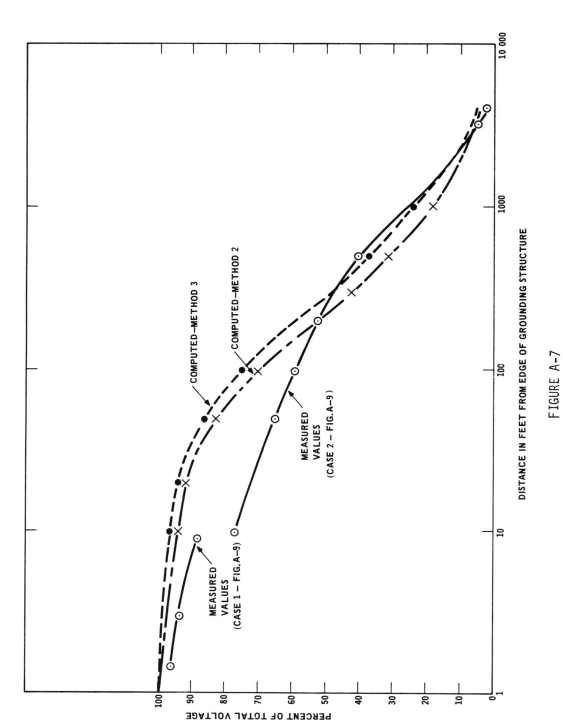

FIGURE A-7

Earth Potential Distribution from the Edge of a Grounding Structure
With Respect to a Remote Earth (Mat Area ≈ 290 000 Square Feet)

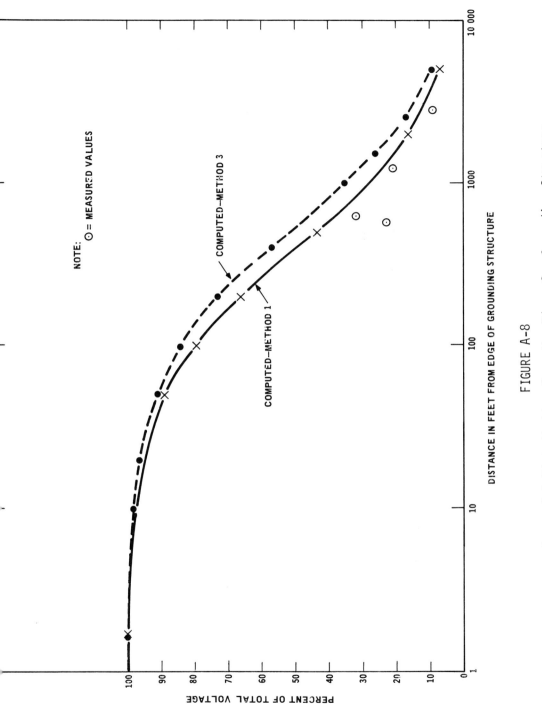

FIGURE A-8

Earth Potential Distribution from the Edge of a Grounding Structure
With Respect to a Remote Earth (Mat Area ≈ 935 000 Square Feet)

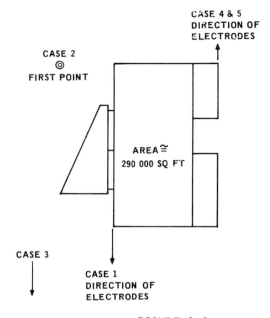

FIGURE A-9

Station Configuration and Location of
Measuring Electrodes - Mat Area 290 000 Square Feet

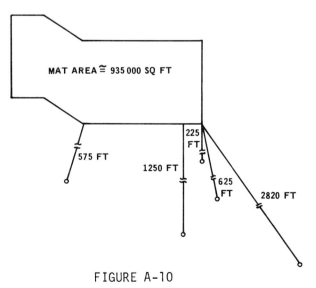

FIGURE A-10

Station Configuration and Location of
Measuring Electrodes - Mat Area 935 000 Square Feet

281

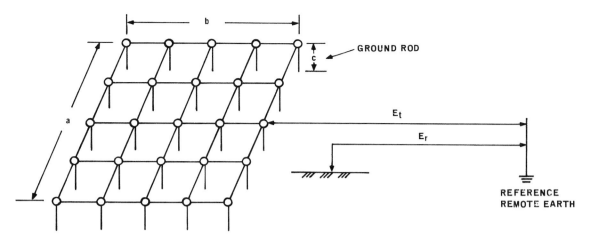

FIGURE A-11

Typical Power Station Ground Mat

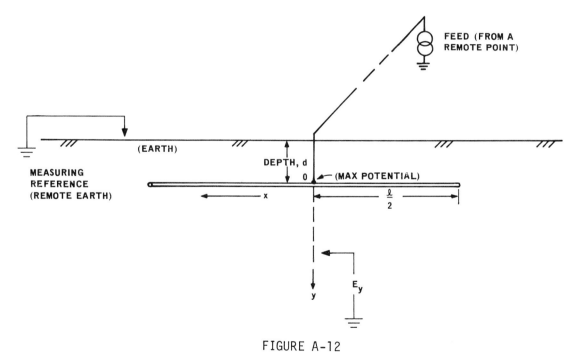

FIGURE A-12

Measurement of Potential Gradient Near a
Buried Horizontal Conductor

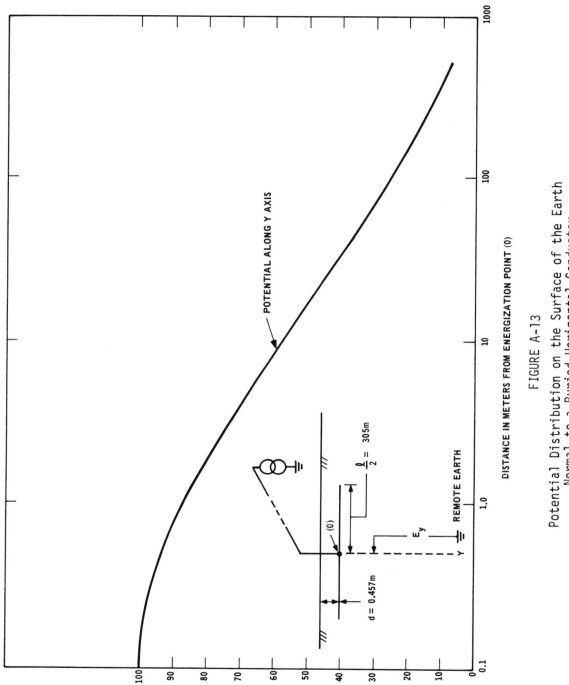

FIGURE A-13

Potential Distribution on the Surface of the Earth
Normal to a Buried Horizontal Conductor

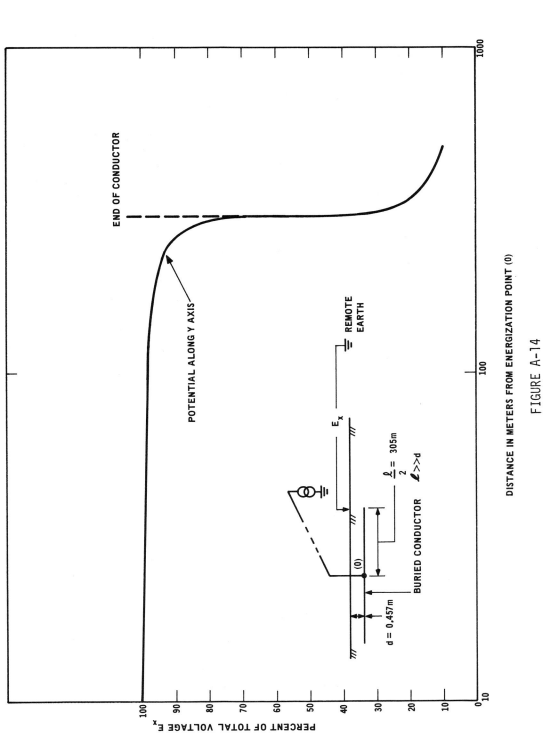

FIGURE A-14

Potential Distribution on the Surface of the Earth
Along a Buried Horizontal Conductor

5.8 APPENDIX B: EARTH POTENTIALS PRODUCED BY MAGNETIC STORMS

B.1 MAGNETIC STORMS - GENERAL INFORMATION

Every magnetic storm produces world-wide effects, but not all
parts of the earth experience disturbances of the same intensity. In
equatorial regions, the effects are least pronounced, while maximum
intensity usually occurs in world-wide auroral lands at distances
between 20° and 30° from the geomagnetic poles. Sometimes, however,
strong fields occur at more southerly points; the auroral belt in
Ontario, for example, is located approximately between latitudes
50° and 60°, and auroral activity occurs in the south polar region.
During massive magnetic storms, the auroral zones tend to shift to
lower latitudes, which exposes a greater proportion of the world's
populous areas to the related earth potential effects.

An auroral band, encircling the earth, acts somewhat like a
single-turn transformer. During a magnetic storm, currents flow in
these bands and induce potential gradients in the earth. The auroral
currents flow in either an east-west or a west-east direction and the
induced earth potential gradients are similarly oriented and are
subject to polarity reversals. Thus radio transmissions oriented in
an east-west or a west-east direction experience maximum interference,
which may last for several hours.

Magnetic storms are generally associated with sunspot activity,
a cyclic phenomenon with an approximate between minima period of 10-11
years. (See Table B-1.)

TABLE B-1
Record of Magnetic Storm Activity

MINIMA	MAXIMA	REMARKS
1934	1940	Exceptionally high intensity
1944	1947	Moderate intensity
1954	1957-58	High intensity
1965	1969-70	Moderate intensity
1975	1979-80	Predicted occurrence

TABLE B-2
Earth Potentials Produced by Magnetic Storms

YEAR	TOTAL P.D. VOLTS	VOLTS/MILE
1938	150 volts	3.0
1938	250 volts	5.0
1940	110 volts	2.2
1940	160 volts	3.2
1941	330 volts	6.6
1941	105 volts	2.1
AVERAGE	184	3.7
MEDIAN	155	3.1

B.2 EARTH POTENTIAL EFFECTS ON POWER AND COMMUNICATION SYSTEMS

The adverse effects on both power and communication facilities produced by an exceptionally intense magnetic storm on March 24, 1940 created considerable concern in engineering circles[3],[4]. The referenced papers only report on the effects observed in the United States, which were most severe in the states adjacent to Canada. Concurrent effects in the southern part of the U.S. were not of sufficient intensity to attract attention. Unfortunately, no one was prepared for such spectacular manifestations of magnetic storms, so none of the affected systems were instrumented to obtain data. During this storm, many carbon protector blocks on long trunk circuits operated and became permanently grounded. Since these protectors typically require from 400 to 600 peak volts to spark over, their operation provided some indication of the intensity of the storm. A telephone trunk circuit between Minneapolis, Minn. and Fargo, N.D. was promptly instrumented, and records were obtained of earth potential difference between these two points during the subsequent storm of March 31, 1940. The distance between grounds was approximately 214 miles, and peaks as high as 340 volts were recorded. This storm was much less intense, as indicated by protector operation; nevertheless potentials of 1.6 V/mi. were recorded.

The Long Lines Dept. of the AT&T Co. conducted an earth potential study on circuits between Philadelphia and Reading, Pa. during 1938 to 1941 inclusive. The distance between grounds was about 50 miles, and the measuring circuit ran approximately east-west.

TABLE B-3

Earth Potential Effects on the Trans-Atlantic and Alaskan Submarine Cables

CABLE	APPROXIMATE DIRECTION	LENGTH (STAT. MILES)	DATE OF RECORDING	PEAK VOLTS	P.D. (V/MI.)	DURATION (HOURS)
Trans-Atlantic	W-E	2300	11 Feb 1958	2650*	1.15	**
Alaskan (Ketchekan-Pt. Angeles)	NW-SE	860	—	130	0.15	3.5
Alaskan (Petersburgh-Wrangell)	NW-SE	35	8 Jul 1958 27 Jul 1958	> 50 >100	>1.43 >2.85	3.5 —

Notes: * With reversals of 700 volts.

        ** System out-of-service for several minutes.

The results of the tests are summarized in Table B-2. Other related data, concerning the Trans-Atlantic cable and the Alaskan submarine cables, are given in Table B-3.

B.3 CONCLUSIONS

The available data concerning the effects of magnetic storms on telephone cable plant is very limited. Consequently, predictions regarding cable exposure tend to be speculative. There appears, however, to be some support for the following conclusions:

1. Magnetic storm intensity reaches a maximum about every 10 to 11 years.

2. Canada is located in proximity to the northern auroral belt and therefore has a relatively high exposure to the effects of magnetic storms. The effects decrease substantially when proceeding south, and are a minimum at the equator.

3. Cables positioned on a north-south line have minimum exposure since they approximately parallel the earth's magnetic field (minimum coupling). Also, it is well established that north-south radio transmission enjoys minimum disturbance from magnetic storms.

4. The closer a cable approaches an east-west orientation, the greater will be its exposure to earth potential differences associated with magnetic storms.

5. The limited data presented here suggests that the magnitudes of earth potential gradients vary inversely with the distance between grounding points. Such a conclusion is quite speculative at this time, but, if true, it is a fortunate relationship from a systems engineering standpoint.

5. 2 V/statute mile has been accepted for some time as the industry design criterion for long trunk circuits and carrier systems of substantial length.

B.4 REFERENCES

1. Dr. V. Gaizauskas, Solar Group, National Research Council, Ottawa.

2. A.G. McNish, "Magnetic Storms". (See Note 1.)

3. W.F. Davidson, "The Magnetic Storm of March 24, 1940 - Effects
    in the Power System". (See Note 2.)

4. L.W. Germaine, "The Magnetic Storm of March 24, 1940 - Effects
    in the Communication System". (See Note 2.)

Notes:

1. Address before the Dinner Meeting of E.E.I. Engineering
    Committees, Chicago, Illinois, May 7, 1940. Reprinted in
    the Edison Electric Institute Bulletin of July, 1940, which is
    available from the library of the National Research Council,
    Ottawa.

2. Probably the best source of information concerning sunspot
    activity and the related effects is Dr. R.R. Heacock,
    Professor of Geophysics, University of Alaska, College, Alaska,
    99735.

6. CORROSION

6.1   INTRODUCTION

In the past century, engineering progress has been largely
dependent upon the development of materials that better withstand
stresses. As the margins of safety were gradually reduced through
improved mathematical techniques, design engineers were able to make
corresponding reductions in initial costs. However, increased
failures due to metal fatigue also began to become evident, directly
confronting engineers with the problem of corrosion.

Metals have always been subject to corrosion, especially in
areas close to salt water. However, rising chemical influences in the
environment in the form of industrial pollution, pesticides, agricultural
fertilizers, and other agents that cause or accelerate corrosion have
intensified the problem. Further complexity has been added by the
increase in stray currents, which can rapidly deteriorate underground
metallic systems. Also, greater use of dissimilar metals will increase
the frequency of galvanic corrosion failures.

Corrosion of communication system plant begins before
installation is complete and continues throughout the useful life of
the plant. Generally, the severity of corrosion may be minimized by
reducing plant exposure, by carefully selecting plant materials, and
by using specific prevention techniques for plant installation and
maintenance.

Corrosion is the result of attack due to natural chemical processes or chemical processes caused by industrial exposures. Of
principal interest in this section are the means for mitigation of
attack on metal cable sheath, other Outside Plant hardware, and other
parts of the network that are exposed to the external environment.

The attack is usually microscopic and always involves the
transfer of electric current between the particular metal structure
and its environment. Lead has been used for cable sheath because it
resists corrosion well and is practical from the standpoint of manufacture and maintenance. However, it is subject to attack in a number
of situations. The recognition of these situations and the treatments
necessary for mitigation of the corrosion will be discussed in the
following sections which concentrate primarily on Outside Plant and
those parts of the network that are directly exposed to the environment.

In addition to lead cable sheath, plastic-covered cables for
local and long distances are now in use. Plastic-covered cables are not
subject to the same attacks that deteriorate lead-covered cables, and
their installation and maintenance are easier. However, although plastic
sheaths reduce corrosion significantly when intact, they usually intensify corrosion attack at splices and other exposed locations.

Outside Plant is usually made of materials that best resist corrosion in natural exposures. Many are made of steel with protective coatings of zinc, while others are made of bronze or aluminum, both of which resist corrosion well in specific situations. However, some situations are encountered in which it may be desirable to apply extra protection.

In industry, in general, the greatest loss due to corrosion results from corrosive attack on steel and iron structures. It is not possible to accurately evaluate this loss. Bridge structures, railroad plant, pipelines, steamships, and a great variety of industrial plant undergo endless corrosive attack. H.H. Uhlig, in an article in "Corrosion"[1] in 1950, estimated this loss to be in excess of six billion dollars per year in the United States alone.

Pipeline companies are particularly concerned about such loss, since they use steel pipes which must be installed in all varieties of soil, some of which are exceedingly corrosive. It is their practice to enclose their metal pipes with insulating coatings, and to apply a cathodic protection system.

Close cooperation between our protection personnel and the personnel of interfering utilities, whether they be pipeline, railroad, power, or other telephone companies, must be maintained to adequately deal with constantly changing conditions.

The purpose of this section is to identify some of the agents and conditions that cause corrosion and to suggest means by which this knowledge may be used to reduce corrosion damage to the communication network. It is hoped that the information presented will contribute to a much needed improvement in corrosion control and protection design.

6.2   THE CORROSION PROCESS

6.2.1   Mechanism

The corrosion process always involves the transfer of electric current between a metal and its environment. This takes place in the form of electrons interacting at two distant areas on the interface of the metal being corroded, and an electrolyte or two electrolytes in electrical relationship with each other. It is because of the various and sometimes very complex ways in which this process can occur that it was so long before the general principle of corrosion was understood.

In the corrosion process metal goes into solution directly at the anode, and, if oxygen is evolved, the solution tends to become acid. At the cathode, hydrogen is usually evolved, which combines with available oxygen and tends to make the solution alkaline. If metal enters solution at the cathode, it does so indirectly by secondary chemical reactions at the metal surface. Metals that are capable of corroding at both the anodic and cathodic surfaces are called amphoteric. Lead and aluminum have amphoteric qualities.

The electrochemical mechanism of corrosion was forecast by Davy in 1825. Speller, in "Corrosion"[2], states that "very little was done on the corrosion problem until after 1900, when the tonnage of iron and steel in use showed rapid increase". Speller also defined the chemical reactions of iron in water in the presence of air, as follows:

$$\underset{\text{metal}}{Fe} \quad + \quad \underset{\text{ions}}{2H^+} \quad \rightarrow \quad \underset{\text{ions}}{Fe^{++}} \quad + \quad \underset{\text{atoms}}{2H} \qquad (1)$$

in which the evolved hydrogen forms an insulating film on the iron surface, slowing down further corrosion.

Either of two actions can occur to remove this film:

a) the formation of water by the hydrogen combining with oxygen from the air:

$$\underset{\text{atoms}}{4H} \quad + \quad \underset{\text{dissolved}}{O_2} \quad \rightarrow 2H_2O \qquad (2a)$$

b) the release of hydrogen as bubbles of gas:

$$\underset{\text{atoms}}{2H} \quad \rightarrow \quad \underset{\text{gas}}{H_2} \qquad (2b)$$

Action (2a) or (2b) permits the reaction to proceed, with the accumulation of Fe^{++} ions in the solution, which are precipitated as rust by the reaction

$$4Fe^{++} \quad + \quad O_2 \quad +2H_2O \quad \rightarrow \quad 4Fe^{+++} \quad + \quad 4OH^- \rightarrow \quad \underset{\text{hydroxide (rust)}}{\text{insoluble ferric}} \qquad (3)$$

The formation of hydrogen film noted in the first reaction is referred to as polarization, and in some instances can completely shut off further action.

Evans and Hoar of the Royal Society in London proved that the reaction occurred at anodic and cathodic areas on partly immersed specimens of plain carbon steel by three different methods. Brown and Mears did the same thing for aluminum. Subsequent work by other researchers proved the same thing for other metals.

The anodic and cathodic areas of these reactions are usually quite close together and occur by reason of differences in the ionization of the solution in direct contact. On the other hand, the two areas can be remote from each other, as in the case of the impressed currents, where anodic areas may be miles from the areas where the stray currents transfer from the earth to the cable sheath or hardware.

6.2.2 Corrosion Rate

The dominant factors in controlling the rate at which corrosion of a metal takes place are listed below.

Factors Chiefly Related to the Metal

A) EFFECTIVE ELECTRODE POTENTIAL OF A METAL IN A SOLUTION:

The initial tendency of a metal to corrode in a solution is measured by the effective electrode potential at any instant between the metal surface and its ions in the solution. This potential is a characteristic of the metal at a specified concentration of its ions, but varies in a definite way as the ion concentration varies. The determination of the specific potential is complex but, from the values that have been established, the relative order of the tendency of some of the commonly used metals to go into solution is as follows (the more active metals are given first and the more noble ones last):

Magnesium, Zinc, Aluminum 2S, Aluminum 12ST, Steel and Iron, Lead-Tin Solders, Lead, Tin, Nickel, Brasses, Copper, Bronze, Silver.

B) OVERVOLTAGE OF HYDROGEN ON THE METAL:

As shown by reaction (1), for each equivalent of metal going into solution (i.e., being corroded) one electrical equivalent must leave it. Hydrogen, however, is involved in the reaction and tends to produce an insulating film that acts as a resistance to the flow of the electric current. The voltage required to overcome this resistance is referred to as the 'overvoltage', since it appears in a voltage measurement between the metal and the solution as an addition to the characteristic electrode potential. In effect, overvoltage operates as the cathode against the evolution of hydrogen and must be overcome if the action is to continue. Bancroft in 'Electrolytic Theory of Corrosion'[3] states 'hydrogen overvoltage is due to polarization by electrically neutral **monoatomic hydrogen, and** hydrogen will be given off as a gas or be dissolved by the solution only as the infinitely small amounts of monatomic hydrogen absorbed on the surface react to form molecular hydrogen'.

C)   CHEMICAL AND PHYSICAL HOMOGENEITY OF THE METAL SURFACE:

The chemical and physical homogeneity of the metal surface affects the rate of corrosion as a result of the variations of ion concentration at the metal interface.

D)   INHERENT ABILITY OF THE METAL TO FORM AN INSOLUBLE PROTECTIVE FILM:

The inherent tendency of metals to form protective films, i.e., to become passive, can slow down or even stop the corrosive action. Stainless steel is such a metal. Similarly, oxygen released at cathodic areas of aluminum produces a thin film of oxide, which also reduces the corrosive action.

Factors That Vary Mainly With the Environment

A)   HYDROGEN-ION ACTIVITY (pH) IN THE SOLUTION:

A very important factor influencing the rate of corrosion of metals, especially of structures in the ground, is the acidity (i.e., the pH) of the water with which they are in contact. Increasing the pH increases the attack on the metal at the interface.

B)   INFLUENCE OF OXYGEN IN SOLUTION IMMEDIATELY IN CONTACT WITH THE METAL:

Many soils are strongly acidic and therefore destructive to metals placed in contact with them. If the metal happens to be in contact with the earth at two sites showing different oxygen ion concentrations, the attack can be severe and may destroy the structure in a short time.

C)   THE SPECIFIC NATURE, CONCENTRATION, AND DISTRIBUTION OF OTHER IONS IN SOLUTION.

D)   THE RATE OF FLOW OF THE SOLUTION IN CONTACT WITH THE METAL:

The motion of the water in contact with the structure can accelerate the rate of corrosion by constantly bringing fresh supplies of ions to the interface. Metal hardware in manholes into which water drains (e.g., after streets have been salted) corrodes increasingly with the passage of each season.

E) THE ABILITY OF THE ENVIRONMENT TO FORM A PROTECTIVE DEPOSIT
ON THE METAL:

Oxygen is also active in solutions that may contact underground
structures. Its ions may attack and destroy protective films
and accelerate the corrosive attack.

F) TEMPERATURE:

Increases of temperature also accelerate the corrosive attack
by causing an increase in the rate of chemical action and by
the production of increased differences of ion concentration
at the interface.

G) CONTACT BETWEEN DISSIMILAR METALS OR OTHER METALS AFFECTING
LOCALIZED CORROSION:

If the underground structure has several metals contacting the
soil, conditions are at once satisfied for galvanic action.
Insulating coatings are resorted to in this situation, but the
integrity of the coatings must be rigorously maintained.

H) STATIC OR CYCLIC STRESSES.

6.3 SOURCES OF CORROSION

6.3.1 Stray Current Corrosion

Stray current corrosion occurs when a stray dc current
that has entered the telephone plant from an external source returns
into the surrounding electrolyte from the cable sheath, manhole
hardware, or ground rods. Corrosion damage will occur at the point
of departure to the extent that one ampere will remove about 70 lb
of lead per year. The chief sources of such current are cathodic
protection systems, direct current powered transit systems, and
mine railways. The current of these dc systems must return to their
source and part of the return path is ground. If there is a cable
in the area the current will use the low resistance cable sheath as
the return path and then leave the sheath or manhole hardware near
the source removing part of the metal with it. The use of PIC
insulated cables greatly reduces sheath damage but increases the
concentration of manhole hardware corrosion.

6.3.2 Galvanic Corrosion

Galvanic corrosion involves two metals forming an electrochemical circuit in the presence of an electrolyte. Variations in
the types of metal, electrolyte, or electric circuit will influence

the rate of corrosion. When the metals are some distance apart they
form a typical galvanic cell. If the metals are in close contact
then localized corrosion (natural cell corrosion) will occur.

The Natural Corrosion Cell is essentially a short-circuited
cell. It may be set up by:

    a) electrodes of two different metals in metallic contact
       and joined by solution covering them both;

    b) electrodes consisting of the same metal surrounded by
       solutions of different concentrations; or

    c) electrodes in the same solution with better access for
       oxygen to one electrode than to the other.

The resistance of the metallic path joining the two areas
that constitute the cathode and anode respectively is negligible.
If the electrolyte connecting the two areas is a concentrated salt
solution, the electrolytic path may also have negligible resistance.
However, in such cases the current is limited by the degree of polarization, which is usually controlled by the limited rate of
replenishment of oxygen to the cathode. Crevice corrosion and
corrosion of soldered joints are common forms of such localized
corrosion.

6.3.3   The pH of the Environment

The Pourbaix diagram is based on the principle that although a metal is likely to corrode over a wide range of pH,
immunity or *passivation* might be attained by controlling the
potential at the interface between the metal and its environment
(The Pourbaix diagram for lead is given in Figure 6-1). Each line
in a Pourbaix diagram represents some electrochemical equilibrium
condition. For example, a horizontal line represents equilibrium
for a reaction involving electrons but not H^+ or OH^- ions. Conversely, a vertical line represents an equilibrium involving H^+ or
or OH^- ions but not electrons. Normal reactions are represented
by lines falling between these extremes.

The Centre Belge D'Etude de la Corrosion (CEBELOR) has
published an atlas[4] containing Poirbaix diagrams for all important
metals, while the principles of the diagrams are enumerated in
Pourbaix's own book *Thermodynamiques des Solutions Equeuses
Deluées*[5] (available in English) or in *Thermodynamics of Dilute
Aqueous Solutions*[6] by Arnold. Pourbaix diagrams are still of a
highly theoretical nature and their present application to practical
situations is as yet limited. However, they exemplify the intensity
of theoretical research in the corrosion field, which hopefully
will lead to much improved prevention techniques.

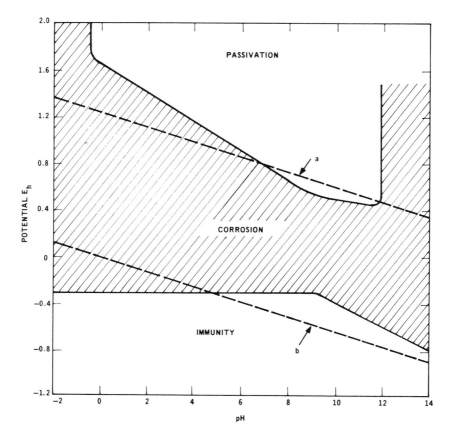

FIGURE 6-1

Pourbaix Diagram for Lead

6.3.4   Corrosive Agents

Brief descriptions of the most important corrosive agents are provided below.

Oxygen

Oxygen is a very strong corrosive agent in situations involving some type of underground plant electrolytic corrosive cell. It causes corrosion by depolarizing the cathodic areas of the sheath. However, under other circumstances, e.g., aerial cables, oxygen may produce a protective effect by forming a continuous oxide film.

Nitrates

Soil is generally low in nitrate content unless contaminated by leakage from sewers or surface drainage. Areas of nitrate corrosion on underground plant are usually black in appearance and covered with loosely adherent finely-divided lead (and antimony, if it is a constituent of the sheath).

Alkalies

Alkali corrosion is usually characterized by the presence
of red tetragonal crystals of lead monoxide. These
crystallize out of saturated solutions of the unstable
alkali plumbites, which are the primary corrosion products
of alkaline attack on lead.

Chlorides

These may have a mildly accelerating influence on sheath
corrosion, due partly to the increase in electrolytic
conductance that they produce and partly to the destructive action of chloride ions on protective films created
by other constituents.

Organic Acids

Acetic acid is the most active corrosive agent of this
group. In the presence of atmospheric CO_2, organic acid
corrosion produces the pigment white lead, which forms
a porous white incrustation over the surface of the
sheath.

Soils

Some soils set up oxygen concentration cells at points of
contact between soil particles and the sheathing. The
activity of these cells depends largely upon the soil
texture. In coarse soils, the corrosive action may cause
performation of the sheath, whereas in fine silty soils
there may be little or no attack.

6.4   CONTROL OF CORROSION

6.4.1   General

The aim of corrosion protection is the retardation or
prevention of direct current flow between metals and their environment. This can be most efficiently and economically achieved if
protection is emphasized at the design stage, where unnecessary
hazards can be forestalled.

Exposed metallic structures should ideally be insulated
from their environment, but in practice it is often neither
feasible nor economical to do so. In such cases, the environmental
hazards must be examined and the most resistant materials selected.
This in itself can create a corrosion problem should a series of
dissimilar metals be coupled together and exposed to the environment. The Bell system is presently adhering to such a policy of
minimum metallic exposures by the use of polyethylene jacketed
cables. However, this increases the stray current action on manhole hardware. More attention must be given to proper coatings
for racking, apparatus cases, and other manhole equipment and to
the use of anodes for manhole grounds.

6.4.2   Control of Stray Current Corrosion

Stray current corrosion can be eliminated by preventing the stray current from entering the telephone plant (e.g., with coatings, insulating joints, etc.), by providing a direct metallic return path with reverse current switch, by applying cathodic protection, or, if the current is small, by draining it off via sacrificial anodes.

To adequately represent the communications industry, communications corrosion engineers should be active in local corrosion coordinating committees. Such organizations can help the engineer to solve existing stray current problems, to stay informed of new installations, and to participate in cooperative tests.

Telephone plant should be checked for stray current by using potential and current measurements as outlined in subsection 6.5. The point of current exit should be located and a direct metallic return path with a reverse current switch should be installed. If the current is small, it may be drained by installing anodes. Under certain conditions it may be necessary to apply a cathodic protection system. For some examples of corrosion problems and their solutions refer to the Proceedings of the 17th Annual Appalachian Underground Corrosion Short Course (A.U.C.S.C.)[7].

In some cases the stray current can be discouraged from entering the plant by placing insulating joint bridged with capacitors in the cable sheath at appropriate locations. These insulating joints will stop both stray and galvanic current flow but may interfere with the electrical protection system. Consequently they should be used with care.

In stray current areas such as Metropolitan Toronto (one of the most severe corrosion areas at present), a combination of cathodic protection and drainage bond wires is used. This involves returning the stray current to its source metallically and draining current from the sheath at locations where bond wires are not economically feasible.

6.4.3   Control of Galvanic Corrosion

Design and Maintenance Engineers should whenever possible avoid earth or electrolytic contacts of dissimilar metals. Useful information in this regard is provided in a report published by the British Service Committee entitled *Corrosion and Its Prevention at Bimetallic Contacts*.[8] The results of the investigation are displayed in a table which indicates for a variety of metals whether the corrosion of one metal is increased after contact with a second. No attempt is made to indicate the degree of corrosion suffered by the first metal in the absence of the second.

Adverse effects at contacts may frequently be avoided by insulating dissimilar metals from each other. This may involve special designs to separate the metals electrically, or, if electrical contact is necessary, the insulation of at least one of the metals. When insulation is functionally impossible or impractical, harmful effects may be avoided in some cases by coating the first metal with either the second metal or a metal chosen for compatibility, or by coating both metals with a compatible metal. Non-metallic coating materials or cathodic protection of the whole joint (see 6.5) are other alternatives. However, in the communications industry, cathodic protection is only applicable to underground plant.

The use of bimetallic contacts should be avoided by any practical means in highly industrialized, high-pollution areas.

6.4.4   Control of Crevice Corrosion

Crevices or pockets where water can collect or rest should be avoided in all structures. Where crevices are inevitable, adequate protective coatings should be applied. For example, zinc coating, provided it is sufficiently thick, is considered satisfactory for steel.

6.4.5   Control of Corrosion at Soldered Joints

The main corrosion hazard introduced by soldering arises from corrosive and hygroscopic residues from the flux. Whenever possible, the use of noncorrosive flux is advisable, e.g., some of the cored solders commercially available. Most of the noncorrosive types contain an activator, which causes the flux residue to decompose at the soldering temperature. Solders containing 25% Indium with 37.5% Pb and 37.5% Sn are particularly effective, since most solders are liable to be attacked by alkali. To avoid the equivalent to a highly localized bimetallic attack, the soldering material for underground connections or joints should be cathodic to the base metal.

6.4.6   Control of Corrosion at Welded Joints

The presence of internal stresses created by heat is known to increase corrosion and corrosion-cracking in several materials. Factors discussed under Control of Corrosion at

Soldered Joints are equally applicable to this problem and perhaps
are the underlying cause of highly localized attacks on loading
coil cases and cable splices.

6.5   CATHODIC PROTECTION

6.5.1   General

This section primarily pertains to corrosion control in
lead cable sheath and is of greater concern to the Corrosion
Engineer than to the Design Engineer.

It is desirable to restrict galvanic current on cable
sheath to avoid metallic contact with foreign plant, particularly
those using copper ground beds for protection. When this is not
possible, insulating joints shunted with capacitors may be used
to break the metallic conducting path for the current.

Alternatively, cathodic protection may be employed. With
this technique, metals below lead in the Normal Potential Series
are used as sacrificial anodes to inhibit corrosion. They are
buried in the immediate vicinity of areas where the cable sheath
is anodic to ground and wire-connected to the cable sheath at the
burial location or in the immediate vicinity. The cable sheath
effectively becomes a cathode, and the expendable sacrificial
anode corrodes.

Should the potentials of the sheath be significantly
large or other cables in the run require similar protection, other
forms of cathodic protection may be necessary. In such cases, more
driving potential than is available from sacrificial anodes is
required to counteract the damaging current. This additional source
can be obtained from a power driven rectifier that directs current
into the soil via expendable anode material buried specifically
for the purpose. The subject of cathodic protection is dealt with
more fully in other reports.[9]

6.5.2   Relation Between Protective Current and Corrosion Current

It is not true that in all circumstances the current that
must be applied to obtain complete cathodic protection is equal to
the corrosion current that would be flowing if there were no
protection. This is demonstrated by Schaschl and Marsh, who have
measured the ratio of protective current to corrosion current in
an extensive series of researches and have collected similar results from the work of others. In neutral or slightly alkaline
liquid or soils, the ratio slightly exceeds unity (1.1 to 1.3 is
usual). For acid liquids, it is usually well below 1.0.

In practice, it may be difficult, if not impossible, to accurately measure corrosion current in underground plant structures. This is usually because there is minimal access for suitable wire connections to the plant. Also, although there are many ways of determining corrosion current, most are highly unreliable because of the many variables involved. The null balance method, however, is **highly** recommended and is discussed in detail later (see 6.6.2).

Potential measurements are an alternative to current measurement for field determinations of whether sufficient current is being applied to give adequate protection. Essentially, these are structure to soil potential measurements using a suitable reference electrode and high sensitivity voltmeters. A saturated copper sulphate electrode is normally used as the reference.

It is generally accepted that corrosion will not occur if the potential of an iron structure is everywhere at least 0.85 volt negative to soil (0.95 volt if sulphate - reducing bacteria are present). To ensure protection for lead, the potential needed is different from that for iron. In the case of lead, Compton, formerly of Bell Labs and now with Miami University, determined **the potential assumed by lead when it is in the soil and in the absence of current. In most soils, this averages about -0.7 volt relative to the $CuSO_4$ saturated electrode.**

6.6   CORROSION MEASUREMENTS

6.6.1   Potential Measurements

Positive structure to soil potentials indicate anodic conditions, while negative potentials indicate cathodic conditions. This may be used to locate points at which corrosion current might be leaving the sheath and causing corrosion on anodic surfaces, and also, if cathodic protection is applied, to determine if the cable is polarized adequately. However, potential measurements require much interpretation, and current flow to or from the cable cannot be determined directly. Positive potentials, for example, are not necessarily indicative of corrosion in action.

Structure to soil potential measurements are not a true indication of the voltage at the structure to soil interface. They are a measure of the potential difference of a cell composed of the cable sheath, the moist earth as an electrolyte, and a reference electrode in contact with soil usually remote from the cable sheath. This means that the contact resistance between the reference electrode and soil must be low, otherwise sizeable meter errors are possible with low sensitivity voltmeters. Also, it is likely that the reference electrode will introduce galvanic potentials, which must be taken into account for scale zero.

In nonstray current areas, potential measurements are a relatively poor means of detecting galvanic action unless the discharge of current is significantly large and highly concentrated on a relatively small isolated section of underground **plant, such as in** rural areas. In stray current areas, it is practically impossible to detect galvanic action because of the overriding effect of much larger stray currents. Should the cable sheath be in intimate contact with backfill containing unburned carbon as found in cinders, potential measurements might be of limited value in both stray and **nonstray** current areas.

In stray current areas, potential measurements are quite useful for the detection of potential gradients, since the currents are significantly large. In heavy stray current areas such as Metro Toronto, the potential measurements may be interpreted directly, if the metering errors mentioned are taken into account. When trouble is suspected, current measurements in cable sheath should be employed to eliminate any uncertainty in the potential measurements.

To ascertain the potential difference, the reference electrode is usually placed on the ground over the structure with one terminal of the voltmeter connected to the structure at some convenient point and the other terminal joined by a wire to the reference electrode. For this measurement, the IR drops in structures can usually be ignored, since metal conductors are in the order of one billion times better as conductors than is average soil. If, when the protecting current is applied, the structure potentials are depressed the specified amounts, then the structure is considered to be adequately protected. If the depth of the structure is small compared to the distance between the structure and the external anode or ground bed, this method may often be sufficiently accurate, but if the depth of structure is considerable, or if the external anode or ground bed is placed close to the structure, there will be an error due to the fact that the equipotential surfaces or IR drops in soil are not perpendicular to the ground.

Even with shallow burial, an error is likely to be introduced if the structure is insulated except at one or two small gaps. The resistances of the bottleneck approaches to these gaps will cause the equipotential surfaces to come together, and any measurement taken on the ground surface will be rendered inaccurate. In all such cases, a margin of safety must be allowed, based preferably on practical experience.

6.6.2   Current Measurements

After cable bonding deficiencies have been corrected and potential measurements have been completed, current measurements in cable sheath are invaluable as a means to pinpoint current discharge locations and to check coverage of cathodic protection systems. Since current measurements may be very laborious and time-consuming, they are not recommended except in emergency situations, such as to determine, with accuracy, the extent of the trouble in a particular solution.

Voltage drop measurements with nominal resistance factors obtained from tables for cable sheath are used by some individuals to obtain a measure of current in the sheath indirectly. However, this method is not recommended. The nominal resistance factors indicated are very unreliable, particularly in a complex system of underground plant. Thus, calculations are required which may introduce additional errors. Also, one cannot assume that underground systems are metallically clear of foreign structures, nor is it likely that cable bonding is 100 percent. Therefore, the total errors may exceed the quantities being measured, and this is usually the case when the current is small (i.e., less than one ampere).

To obtain current measurements in cable sheath directly in the case of serious trouble, or if accuracy is important, the zero resistance or null balance method (Figure 6-2) is recommended.

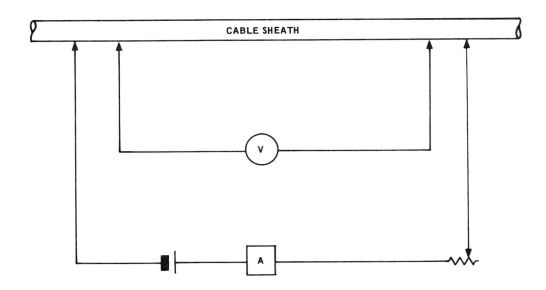

FIGURE 6-2

Zero Resistance - Null Balance
Current Measurement

In areas highly conducive to stray current corrosion, current direction is usually sufficient, and voltage drop measurements in the sheath are adequate.

With the equipment presently available, the zero resistance method can be quite accurate, as currents less than 10 milliamperes may be measured with less than one percent error. The only obstacle is ready access to the cable where small currents are likely to be encountered, as is often the case in rural areas. As the stray current problem extends into the rural areas, this may be sufficient reason to bring test leads, permanently connected to cable sheath at regular intervals, to a surface of convenience. This is done in the LD-4 system.

6.7   MATERIALS

Cheap metallic alloys are most widely used for commercial applications today. However, such alloys are used only sparingly in outside plant facilities in comparison to lead and iron products, since they are extremely susceptible to corrosion. If the serious corrosion problems associated with these commercial alloys can be overcome, they would be more suitable for use since they are more economical.

Lead has been the most widely used metal in outside plant in the past because of its resistance to corrosion. However, pollution of soil by direct current from grounded dc systems now seriously threatens its viability. Precautions are therefore required if lead is still to be used in communications systems.

Also of importance is the selection of the most suitable materials, i.e., the most corrosion-resistant metals for the particular environment. The design engineer must consult appropriate guidelines when making this selection.[12,13] Finally, it is necessary that work proceed towards the identification of the most suitable alloys and the development of better coating materials.

6.8   ECONOMIC AND ENVIRONMENTAL CONSIDERATIONS

6.8.1   Corrosion Mitigation

It is apparent that losses caused by corrosion in the plant that makes up the communication network must receive major attention in each of the design, construction, and maintenance phases.

Fortunately, designers of Outside Plant realize their
responsibility and take advantage of economically viable designs
that minimize the threat of corrosion. At the construction stage,
existing Practices provide some guidelines for corrosion control.

It is in the maintenance phase where the greatest errors
in judgement and operation can occur. Although Practices exist
to guide individuals, plant is constantly undergoing variations
in corrosion exposure - natural changes resulting from the weather,
pollution, and changes to equipment made by power companies, rail-
roads, or pipeline companies can all have a marked effect on
plant. Such changes occur constantly, and the protection measures
for plant must be adjusted in step.

6.8.2   Future Environmental Hazards

About 10 years ago, prominent engineers and scientists
suggested that practically all corrosion of underground metallic
structures was due to the Natural Corrosion Process and that stray
current electrolysis, caused by external forces, was a thing of the
past. More recent experience shows that stray current continues
to be the predominant cause of corrosion of lead cable sheath used
by Bell Canada. Certainly the records in the Toronto area over
the last 16 years bear this out, as stray current from grounded
direct current systems accounted for practically all corrosion of
the lead cable sheath. It is expected that Hydro dc transmission
lines will eventually increase the problem, since it is now economical to transmit power by this means over long distances. Up
to now, the main current sources have been electrical railways
and cathodic protection systems that still are expanding and
increasing in number. Present plans for introducing electric
transit systems into several larger Canadian cities will intensify
this problem. There is also a trend for industry to set up plants
in more remote locations, which will have a serious effect on the
environment.

Therefore, as underground plant systems continue to grow,
failure to keep pace with the stray current problem, which affects
all systems, could prove disastrous. It is also felt that the hazards
created by alternating currents could be a significant corrosion
problem in the future.

6.9   CONCLUSIONS

It is difficult to estimate with any accuracy the long-
term effects of corrosion on either service or costs in the communication network. Several attempts have been made to evaluate the
present and future effects of corrosion on the communication network, without conclusive results. However, it is clear that
the environment is becoming more hostile, with a greater concentration of utilities that have an interaction with each other and more
chemical pollution that attacks the communication network.

Much has been done to reduce corrosion by the use of non-metallic materials. One very effective use of nonmetallics has been the plastic covering of cables, which drastically reduces corrosion while the envelope is intact. There is still concern where the bare metal is exposed in both aerial and underground cable installations, especially in chemical pollution, salt spray, and stray current environments. With a trend towards ready access terminals, more hardware that is exposed to the environment, and more sensitive circuits, all with higher reliability requirements, corrosion is becoming a more important factor in planning the communication network.

To adequately mitigate the harmful effects of corrosion requires a systematic approach that involves the environmental constraints, component design, and the operation and maintenance of the network. Adequate corrosion protection requires a good working knowledge of corrosion within the communication industry and a program to ensure that corrosion control is considered in the design stage, during installation, and in the subsequent maintenance of plant. There is evidence to indicate that corrosion is at this time not being dealt with properly in many phases of communication planning and operations. It is expected that more corrosion problems will become evident in the near future, which will result in much higher maintenance costs, degradation of service, and the retirement of many units or plants before their economic life cycle is completed.

The design engineer must select materials that will in no way interfere with the corrosion process of lead, which is the primary exposed metal in the communication industry. This means that there is practically no choice but to completely insulate dissimilar metals from the environment. This is most important where metals other than lead are in bimetallic contact for protection reasons. If this is not practical or economically feasible, the use of expendable anodes or other means of providing corrosion protection should be considered for underground plant. Lead-aluminum and iron-aluminum couples are particularly vulnerable. There are many handbooks and references available to guide design, development, and operating engineers. It is apparent, however, that there is very little quantitative information on the problems associated with corrosion. This makes it difficult to evaluate the extent of corrosive damage or the nature of this damage for improvement in design. It is therefore desirable that efforts be made to record and utilize information concerning the effect of corrosion on the operation of the communication network, and that texts should be augmented by field experience and an active protection program. There are also some basic Practices available on corrosion, insulating joints, bonding, and grounding that should be fully utilized.

It is recommended that a better overall system approach be used in the mitigation of corrosion, along with a better use of present Practices and training to acquaint designers, and operating and maintenance personnel, with corrosion and its control.

6.10   REFERENCES

1.   Uhlig, "Corrosion", 1950.

2.   F.N. Speller, "Corrosion".

3.   Bancroft, "Electrolytic Theory of Corrosion".

4.   Centre Belge D'Etude de la Corrosion,"CEBELOR Atlas".

5.   Pourbaix,"Thermodynamiques des Solutions Equeuses Déluées".

6.   Arnold,"Thermodynamics of Dilute Aqueous Solutions".

7.   Proceedings of the 17th Annual Appalachian Underground Corrosion Short Course 1972.

8.   British Service Committee, "Corrosion and its Prevention at Bimetallic Contacts".

9.   Designing Impressed Current Cathodic Protection Systems with Durco Anodes, *The Duriron Co.*, 2nd Ed., 1970.

10.   "Electrochemical Corrosion of Underground Metallic Structures, CSA Code, Part III, C22.3, No. 4.

11.   H.H. Uhlig, "Corrosion and Corrosion Control", *Wiley*, 2nd Ed. 1971.

12.   Francis L. Laque and H.R. Copson, "Corrosion Resistance of Metals and Alloys", *Van Nostrand-Rein*, 2nd Ed.

13.   Ibert Mellan, "Corrosion Resistant Materials Handbook", *Noyes*.

14.   H.H. Uhlig, "The Corrosion Handbook", *Wiley*, 1948.

15.   "Corrosion and Its Prevention" (Introduction by U.R. Evans and V.E. Rance, *H.M. Stationery office*).

16.   International Telegraph and Telephone Consultative Committee (CCITT), White Book, Vol. IX, Parts 1 and 2.

17.   The National Association of Corrosion Engineers (NACE), text for Basic Corrosion Course.

18.   Circular 570, *U.S. Bureau of Standards*.

7. OVER-VOLTAGE IN AC POWER UTILIZATION CIRCUITS
(600 V rms and Less)

7.1 INTRODUCTION

Communication operating companies have recognized for many years that telecommunication apparatus, radio facilities, tower lighting, and electronic equipment powered from external sources can be damaged by over-voltage surges originating on primary and secondary power distribution circuits. Some measurements, but chiefly experience, have established that these over-voltage surges do attain magnitudes many times greater than normal operating voltages. They are inherently associated with the operation of a power distribution network and cannot be prevented by practical means. Sources of over-voltage surges are many and varied; some of the more important of them are listed below.

a) Lightning strokes to primary and secondary distribution lines.

b) Arcing line faults.

c) Switching transients.

d) Operation of current limiting fuses.

e) Ferroresonance.

f) Feedback of surges generated within load complexes, which then enter electrically adjacent equipment.

g) Induction in low voltage circuits through mutual coupling with high voltage disturbing sources.

AC-powered equipment is typically designed to withstand normal over-voltage conditions such as line regulation, surges associated with the energizing and de-energizing of a specific unit, and other moderate switching surges. This appears to be an economical practice, because protection against excessive voltages can be provided for an entire installation by the application of a properly selected and installed power type arrester. Equipment designers must recognize, however, the existence of external hazards and establish design criteria that assure a withstand capability compatible with the anticipated protection measures to be employed in the field.

> NOTE: *The provision of adequate protection to assure service continuity and to minimize maintenance expense is the responsibility of the user.*

7.2   SOURCES OF INFORMATION

Exposure of users' wiring and equipment to over-voltage surges has been known within the power utility industry. Investigations conducted by the power industry and manufacturers of transmission apparatus have been directed chiefly toward the effects on distribution transformers and their protection against surges on both primary and secondary circuits.

A few years ago, a working group was established within the IEEE Power Group to investigate the problem of surges in ac power circuits of 600V rms and less. To date, two useful documents have resulted[1,2]. This working group is proceeding with a field study of surge magnitudes using peak voltage recorders, but the results of the study have not yet been published.

7.3   EXPOSURE

It is generally accepted that the larger magnitude surges appearing in secondary and branch load circuits are the result of lightning striking either the neutral or a phase conductor of primary distribution lines. A direct stroke to a neutral can elevate the potential of this conductor and in turn raise the potential of secondary neutrals, especially in the vicinity of the stroke point. Two cases will be considered.

CASE 1 - Strokes that are either sufficiently remote from a distribution transformer or of such a magnitude that they do not operate the arrester associated with the distribution transformer. For this example, consider a 9 kV rated arrester, which has a typical sparkover voltage on a 1-1/2 × 40 wave of 36 kV crest. Thus, from a stroke to the phase conductor, a surge voltage approaching 36 kV crest could appear across the primary winding. This surge can now propagate from primary to secondary through the electrostatic and magnetic coupling of the transformer. The electrostatic components will attain maximum values on open or lightly loaded circuits and appear to range from 0.05 to 0.1 times the sparkover value of the primary arrester (in this example roughly 1.8 to 3.6 kV peak). The duration, however, will be quite short — not exceeding a few microseconds. This may appear to be a substantial over-voltage to appear in a nominal 120 V circuit. However, because of its short duration and attenuation before reaching connected loads, it is not believed to present a serious hazard. Investigators generally agree that surges will transfer electromagnetically from primary to secondary approximately in accordance with the transformer turns ratio. Thus, for a 13200/7620 line, the step-down ratio would be 63.5/1. In such a case, the magnetically coupled surge component would not exceed about 567 peak volts. A 9 kV rated arrester with an external gap will have a sparkover voltage of 43 kV, and a surge in the 120 volt winding could approach 667 peak volts. Such voltages ordinarily should present no problems.

CASE 2 - Lightning surges on phase conductors of sufficient magnitudes to operate primary arresters, or direct strokes to neutral conductors. In such a case, a sizeable portion of the surge current will enter the ground at the transformer pole, and a high potential will appear at the secondary neutral tap of the transformer. From here, surges will flow over the service conductors and distribute in load circuitry that is at a lower potential with respect to ground. It is this type of situation that investigators consider to be the major source of surges which damage utilization equipment. Low voltage wiring is nominally rated at 600 V rms, but will typically withstand surge voltages of 5.0 to 6.0 kV. It is the more vulnerable equipment connected to so-called low voltage wiring that presents the problem.

To obtain some concept of the exposure problem, a simple situation will be considered briefly that might relate to any small load such as a remote land-based radio station or a rural central office. The model selected for consideration is shown in Figure 7-1. The inductance of 30 feet of conductor plus the ground rod resistance (100 ohms in 250 ohms-metre soil) presents a high impedance to ground for lightning surge current. As a result, high surge potential will appear on the secondary neutral tap of the distribution transformer, and surge currents will flow in the system as indicated in Figure 7-1. The magnitude of surges appearing at connected utilization equipment will depend on the relative surge impedances throughout the system. It is evident, therefore, that surge voltages appearing across low impedance loads will be substantially less than those appearing across higher impedance loads, open switches, etc. Characterization by analytical means would lead to uncertain results because of the difficulty of evaluating circuit variables.

What is needed for engineering purposes is measured statistical data concerning the probability distribution of surge magnitudes and waveshapes that appear at commonly used types of

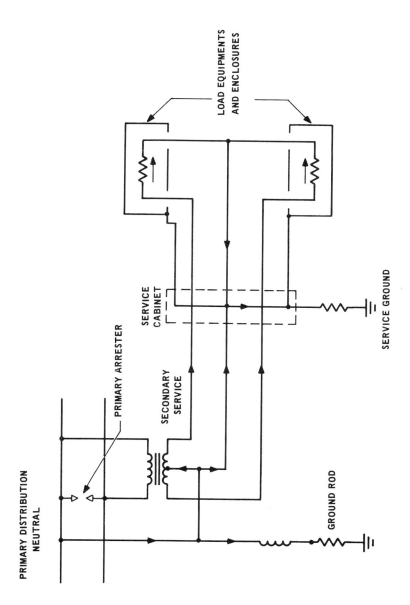

PRIMARY DISTRIBUTION
NEUTRAL

PRIMARY ARRESTER

SECONDARY
SERVICE

SERVICE
CABINET

LOAD EQUIPMENTS
AND ENCLOSURES

SERVICE GROUND

GROUND ROD

FIGURE 7-1

Lightning Surges in AC Power Utilization Circuitry

load equipment in a variety of installations. Such an investigation
is in progress, but quantitative studies of how surges distribute in
complex power distribution and utilization systems do not attract
much interest at present because of the expense involved and the
availability of effective and relatively inexpensive protection
measures.

7.4   SURGE MAGNITUDES

Reference 1 presents the most comprehensive data published to
date. Although the instrumentation used and other aspects of the
studies introduced serious limitations, the results are nevertheless
useful and constitute a valuable contribution to a little-explored
subject. The authors point out that, in this preliminary exploration,
no consideration was given to local geographical and meteorological
conditions in locating recording equipment. It appears that about
50 unipolarity single stage peak voltage recorders were used plus
one cathode ray oscillograph. Four hundred locations in 20 cities
were monitored for short periods of time at any one location (the
entire period of the study was 2 years). At a majority of the
locations, no significant data was obtained other than switching
surges, probably because of the generally urban nature of the
environment. However, the following summary presented by the investigators contains some interesting generalized engineering
information.

   a) Surges due to lightning reached magnitudes of 5600 volts.

   b) Internal generated surges as high as 2500 volts were
       recorded.

   c) In industrial circuits, the level of surges are lower
       than in residential circuits. However, switching surges
       on the load side of the switch can be severe.

In Figure 7-2, a rough attempt has been made to present the very
limited magnitude data obtained in these studies. Since unipolarity
recorders were used, it should be recognized that in some cases the
surge voltages recorded may have been the second half-cycle of an
oscillatory wave. Also, since the surges recorded by the scope are
oscillatory with a rapid decrement, it is possible that several
higher magnitude voltages were not indicated by the voltage recorders.
Recent data obtained by BTL at Chester, N.J., over a period of
26 months (which includes the 1971 and 1972 lightning seasons) are
shown in Table 7-1.

Some unpublished amplitude information was obtained several
years ago by the Bell Telephone Laboratories with a four stage
bi-polar recorder connected to a branch circuit in one of the
buildings at the Chester, N.J. Laboratory. These data, based on four
storms during a period of observation of approximately one month,
are shown in Table 7-2.

TABLE 7-1

Magnitude of Surges on 120 V Branch Circuits
(BTL Measurements, Location 1)

NUMBER OF MEASUREMENTS	PEAK VOLTAGE RANGE
98	300-500
14	500-1000
1	1000-1500
3	> 1500

TABLE 7-2

Magnitude of Surges
(Martzloff's Measurements)

NUMBER OF MEASUREMENTS	HIGHEST PEAK VOLTAGE RECORDED
22	450
8	750
1	1250

An automatic recording CRO was also connected to a branch circuit at another laboratory location in Chester during the same period. The waveshapes obtained are given in subsection 7.5, and the amplitudes are shown in Table 7-3.

TABLE 7-3

Magnitude of Surges on 120 V Branch Circuits
(BTL Measurements, Location 2)

NUMBER OF MEASUREMENTS	PEAK VOLTS
2	350
1	800

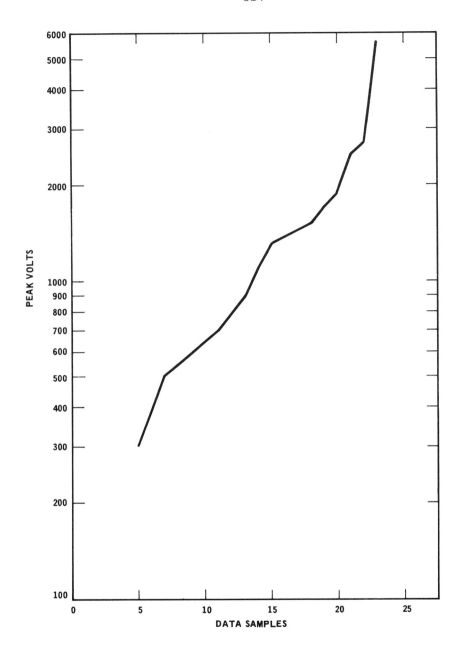

FIGURE 7-2

Peak Voltage Measurements on Low
Voltage ac Branch Circuits (Reference 1)

7.5   SURGE WAVESHAPES

Some brief waveshape information is given in the Martzloff-Hahn paper[2]. Five oscillograms are shown, but the reproduction is poor. About all that can be generalized from these oscillograms is that:

a)   at the service entrance to a home, the wave is oscillatory, and the major voltage components occur in the first 5 to 10 microseconds, and

b)   switching surges are less oscillatory and persist for much longer periods (> 100 μs).

Much better records were obtained in the brief BTL study previously mentioned, a typical example of which is given in Figure 7-3. There is correlation to the extent that the waves are oscillatory and the major voltage component lasts only about 10 microseconds.

7.6   CORRELATION OF MEASURED DATA WITH EXPERIENCE

Many years ago, the writer was confronted with a rash of failures in ac powered land based mobile radio receivers. These were vacuum tube devices employing 117 volt rectifier tubes which were damaged during thunderstorms. Laboratory investigation showed that, with a 1-1/2 × 40 impulse, surges of about 5 kV crest were required to reproduce the type of damage experienced in the field. The application of nominally rated 2000 volt secondary type arresters eliminated the damage to the radio apparatus and also prevented burn-out of lamps in the tower lighting.

During the initial trial of TJ radio, which employs semi-conductor rectifier bridges, the above simple protection measure was not effective. In this case, laboratory investigation indicated that surges in the range of 800 to 1000 peak volts would damage the bridge. Damage was experienced on perfectly clear days, only a few hours after being powered from the ac line. The cause of trouble was switching transients.

More recently, rectifier bridge damage was experienced in large L-carrier powering points. Field observations were made at one such location in Indiana for about one week during the winter. These tests were confined to observing the nature of surges generated by switching equipment within the installation. It was confirmed that such transients can have substantial amplitudes and relatively long durations, somewhat similar to the illustration given in Reference 2, which occurred from oil burner ignition. Figures 7-4 to 7-7 show some of the higher-amplitude switching transients obtained in the BTL study. It should be noted that a filter input was employed to block 60 Hz and 180 Hz voltages. Therefore, in practice, total

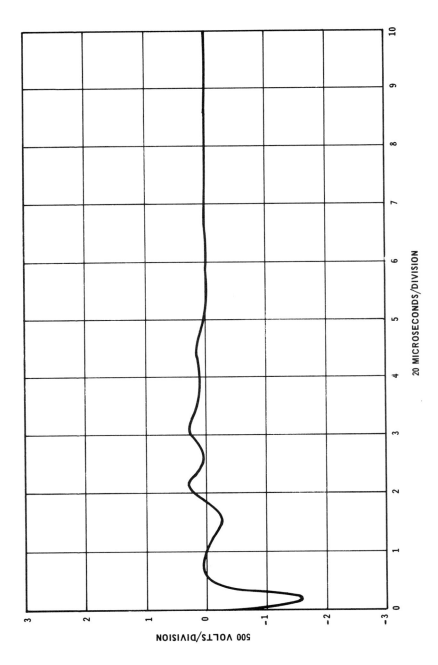

FIGURE 7-3

Phase to Neutral Lightning Surge on a 120 V (rms) Branch Circuit

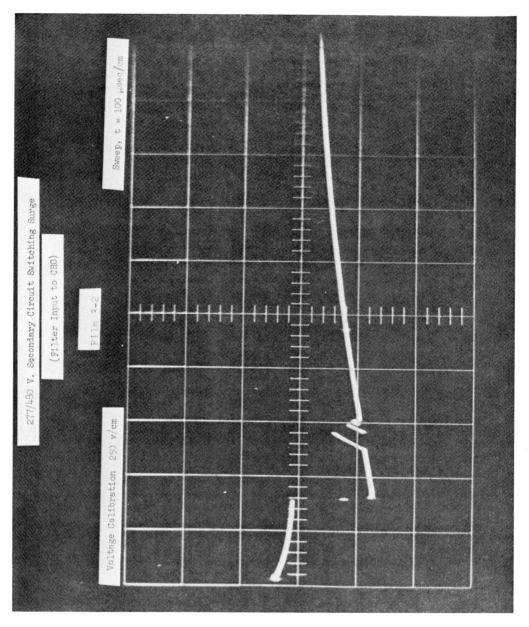

FIGURE 7-4

277/480 V Secondary Circuit Switching Surge
Trace 1: 250 V/cm, 100 µs/cm

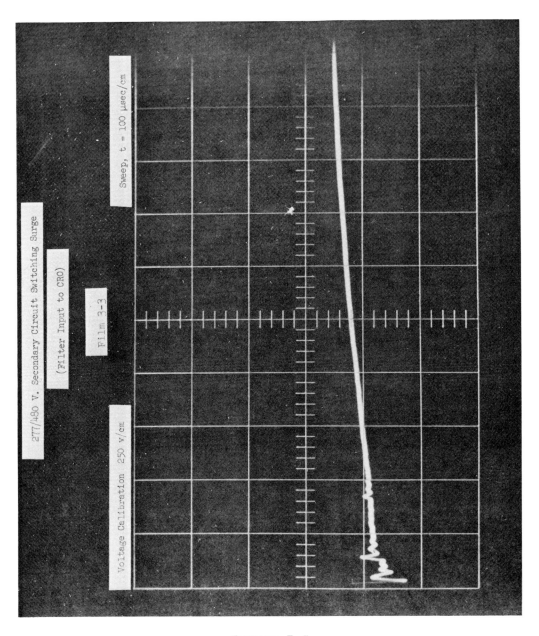

FIGURE 7-5

277/480 V Secondary Circuit Switching Surge
Trace 2:   250 V/cm, 100 µs/cm

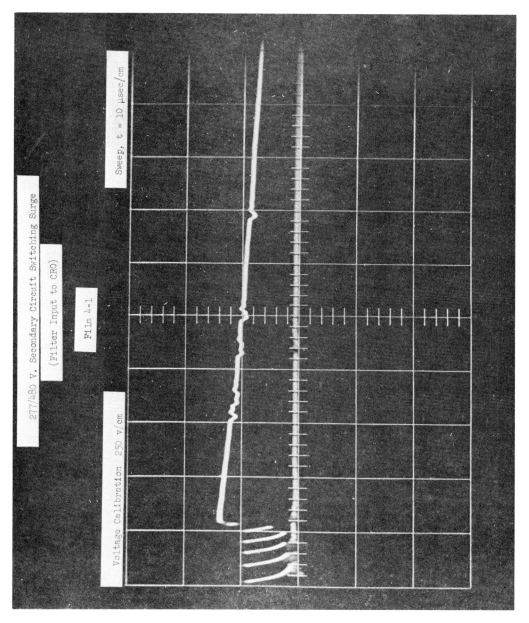

FIGURE 7-6

277/480 V Secondary Circuit Switching Surge
Trace 3: 250 V/cm, 10 μs/cm

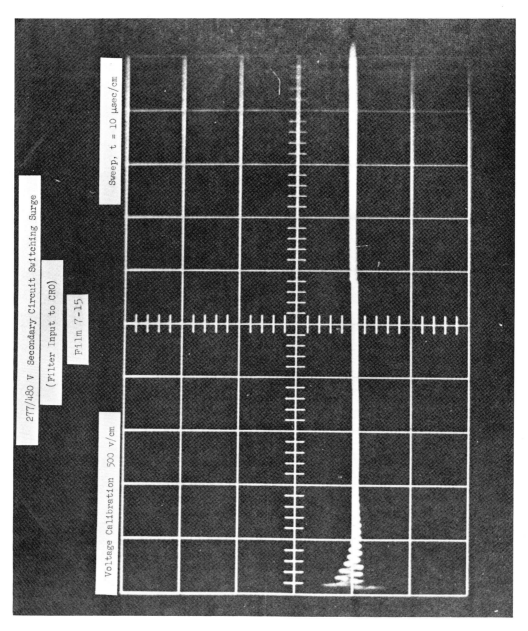

FIGURE 7-7

277/480 V Secondary Circuit Switching Surge
Trace 4: 500 V/cm, 10 µs/cm

amplitudes would be substantially greater if surges appeared at the cresting of the fundamental power voltage. Probably the most interesting record obtained was observed across the primary side of the rectifier input transformer when energized by closing of its breaker (Figure 7-7). The maximum excursion of the initial peak may have been about 1000 volts if the scope had been set to resolve such a short time event. However, rectifier damage did not result.

7.7 REFERENCES

1.   F.D. Martzloff, G.J. Hahn, "Surge Voltages in Residential Power Circuits", *IEEE* Transactions.

2.   IEEE Working Group (Surge Protective Devices Committee), "Bibliography – Surge Voltages in AC Power Circuits Rated 600 Volts and Less", *IEEE* Transactions Paper 69 TP 620-PWR, June 1969.

8. ELECTROMAGNETIC PULSE: EFFECTS ON COMMUNICATION SYSTEMS AND PROTECTION POLICY

8.1 INTRODUCTION

Electromagnetic Pulse (EMP) occurs when a nuclear explosion at or near the surface of the earth causes an electromagnetic field to propagate outward from the point of explosion. Depending on its intensity, EMP can temporarily block or permanently damage components of a communication system. Solid state devices and magnetic memories are particularly susceptible. The degree to which communication facilities are exposed to EMP effects depends on such factors as the type and size of the explosive device, its position with respect to earth, and the proximity of communication plant to 'point zero'.

EMP protection engineering requires the following kind of information, presently available only from restricted sources:

1) field strength levels that specific systems should withstand,

2) numerical values of critical parameters,

3) justification of the cost of special protection measures.

The problem of possible exposure to EMP is distinctly different from the hazards of lightning and low frequency induction. The latter are predictable sources of interference and protection decisions can be based on past experience and substantial amounts of available test data. However, in the EMP field, no such guide is available to the communication engineer. Information regarding EMP protection policy remains classified and can only be obtained through official channels after both establishing a need and arranging appropriate clearances. Unclassified material is too general to provide a satisfactory basis for establishing system protection requirements.

8.2 PROTECTION CONSIDERATIONS

It is generally believed that the effects of EMP on telephone plant are similar to those of lightning, since they both produce heavy surge currents in the metallic shields of cables and induce disturbing currents in apparatus circuitry. Consequently, it is suggested in unclassified EMP literature that a lightning stroke to ground can be used as a rough model for simulating EMP effects. However, little is given concerning the scaling of lightning data to obtain representative levels of EMP.

EMP protection requirements are primarily determined by national defence strategy. The level of system protection above that ordinarily provided for lightning and low frequency induction and the associated expenditure are dictated by national defence policy.

Classified information obviously cannot be discussed in this text, and there seems to be little benefit in reviewing unclassified material that is already available in documents of the type referenced in Subsection 8.3. It is important to note that EMP concepts have changed significantly over the years, and decisions regarding the EMP status of new systems should only be made after a review of up-dated technical information and current defence policy.

At present, the position of the Canadian Government regarding EMP is to depend on redundancy of communication facilities rather than to resort to protection measures other than those normally provided for protection against lightning and power induction.

8.3 UNCLASSIFIED SOURCES OF EMP AND PHYSICAL HARDENING INFORMATION

1. Hardening of Structures Against Electromagnetic Pulse, *AT&T Co.*, PEL 7314, May 19, 1964.

2. Outside Plant Hardening, *AT&T Co.*, PEL 6645, April 12, 1960.

3. Bell System Practice, Section 760-150-001.

4. Effects of Nuclear Weapons, Superintendent of Documents, *U.S. Government Printing Office*, Washington 25, D.C.

   (This document provides background material on the physical effects of nuclear explosions.)

5. EMP Protection for Emergency Operation Centers, Document TR 16A, U.S. *U.S. Government Printing Office*, No. 412-509, May 1971.

6. Jack Bridges, Notes from EMP Awareness Course, *Illinois Institute of Technology*.

   NOTE: Only information from recognized sources, preferably published after 1963, should be considered for engineering EMP protection.